Introducing the Dragonflies
of British Columbia and the Yukon

Robert A. Cannings

Victoria, Canada

Copyright 2002 by the Royal British Columbia Museum.

Published by the Royal British Columbia Museum, 675 Belleville Street, Victoria, British Columbia, V8W 9W2, Canada.

All rights reserved. No part of this book may be reproduced or transmitted in any form by any means without permission in writing from the publisher, except by a reviewer, who may quote brief passages in a review.

Funding for this research and extension project was provided by Forest Renewal BC. (Funding assistance by Forest Renewal BC does not imply endorsement of any statements or information contained herein.)

All drawings by Robert A. Cannings (© RBCM).
Cover photograph by Robert A. Cannings (© R.A.C.).
Back cover photographs by George Doerksen and Robert A. Cannings (all © RBCM).
Credit and copyright information for interior photographs on p. 94.

Edited, designed and typeset by Gerry Truscott, RBCM.
Typeset in Palatino 9.5/11 (body) and Optima 9/11 (captions).
Cover design by Chris Tyrrell, RBCM.
FRBC funding coordinated by Andrew Harcombe, Ministry of Sustainable Resource Management.

Second printing: 2005; third printing: 2019.

National Library of Canada Cataloguing in Publication Data
Cannings, Robert A., 1948-
   Introducing the dragonflies of British Columbia and the Yukon

   Includes bibliographical references.
   ISBN 0-7726-4637-6

   1. Dragonflies – British Columbia. 2. Dragonflies – Yukon Territory. I. Royal British Columbia Museum.

   QL520.2.C3C36 2002   595.7'33'09711   2001-960263-4

Contents

Introducing Dragonflies 5
Life History and Behaviour 9
Dragonfly Habitats in B.C. and the Yukon 14
Studying Dragonflies 18
Dragonfly Conservation 19

How to Find and Identify Dragonflies 20
The Dragonfly Body 21
Key to Dragonfly Families in Our Region 23

The Dragonflies of British Columbia and the Yukon 25
Range Types 26
Our Region (map) 27
Damselflies (Suborder Zygoptera) 28
Jewelwings (Family Calopterygidae) 28
Spreadwings (Family Lestidae) 29
Pond Damsels (Family Coenagrionidae) 32
Dragonflies (Suborder Anisoptera) 43
Petaltails (Family Petaluridae) 43
Darners (Family Aeshnidae) 43
Clubtails (Family Gomphidae) 53
Spiketails (Family Cordulegastridae) 57
Cruisers (Family Macromiidae) 58
Emeralds (Family Corduliidae) 59
Skimmers (Family Libellulidae) 71

Glossary 90
Recommended Reading 93
Acknowledgements 94
Index 95

For my nephew, Russell Cannings, another watcher at the pond.

A male Cardinal Meadowhawk – for some, seeing a dragonfly up close can be a life-changing experience.

Today I saw the dragon-fly
Come from the wells where he did lie.

An inner impulse rent the veil
Of his old husk: from head to tail
Came out clear plates of sapphire mail.

He dried his wings: like gauze they grew:
Thro' crofts and pastures wet with dew
A living flash of light he flew.

– Alfred Lord Tennyson,
from *The Two Voices*

Introducing Dragonflies

With a quick whisper and blur of wings, he appeared before me as I sat by his favourite perch. Hovering fiery-red in the spring sun, he examined this giant newcomer to his territory. Then he settled on the twig, inches from my face, cocking his head this way and that, the huge spangled eyes alert to any movement. Suddenly, he was gone, only to return a moment later with a long-legged fly in his jaws. I was so close and the air so still that I could hear the fly's body crackling as it disappeared between the dragonfly's jaws.

Growing up in the Okanagan Valley, I had regularly visited the local ponds searching for turtles, salamanders and backswimmers. And I loved dragonflies, flashing in the sunlight. But I didn't really begin watching them until much later, that day at a Vancouver pond, when that stunning red Cardinal Meadowhawk flew into my life. The experience changed me. I was attending university, seriously studying insects – but from that moment on, I began looking at dragonflies carefully and have never stopped.

Dragonflies are astonishing animals. They have fascinated people all over the world and down through the ages with their bright colours and dashing flight – poets have written about them, artists have painted them, and children have marvelled at them from the edge of a pond. In the last decade, more and more people who love the outdoors have become interested in dragonflies. This has sparked an increase in popular books on dragonflies and a proliferation of internet sites dealing with these insects. The popularity of watching and studying dragonflies seems to be on the rise.

Most insects are small and difficult to see. There are so many different kinds that it's hard to get to know even a few groups of them very well, even for a biologist. But insects are the most abundant organisms on earth, and they are critically important to the well-being of our environment. We would do well to learn more about them – and dragonflies are an excellent group to start with.

Introducing Dragonflies

Male Canada Darner. Big eyes, strong jaws, spiny legs and powerful wings make a superb predator.

A patient observer can easily watch dragonflies going about their lives. They are large insects, colourful and easy to find, if you know where to look. As adults, they are active by day. We know of 88 species living in British Columbia and the Yukon in ten rather distinctive families – this is a group small enough to get to know with just a bit of effort. With a little help and practice, many of them can be identified in the field.

For many years, I have used dragonflies to introduce both children and adults to the complex and exciting world of insects. Meeting these colourful flyers up close helps us understand them and their prominence in freshwater ecosystems. We grow to respect them and all the other small creatures we share the Earth with. Perhaps with this book in hand you will become as intrigued by dragonflies as I am, and head out to the water's edge to learn their names and habits.

The insect order Odonata (Greek for "toothed jaws") contains the groups of insects known in English as dragonflies and damselflies, but in this book I use the name "dragonflies" to refer to the whole order. "Odonates" is another name that is gaining popularity. The Odonata contains about 5,500 named species in 33 families worldwide. For comparison, there are roughly the same number of mammal species in the world and almost twice as many birds. Most dragonflies live in the tropics, but a few have adapted to the cooler temperatures of higher latitudes; even in our region there are many more species living in the south than the north.

Dragonflies are among the most ancient insects – their ancestral line goes back to the Carboniferous Period, about 300 to 350 million years ago. They have retained many primitive characteristics and developed some specialized features for a successful aerial and predatory lifestyle. Dragonflies share with mayflies the ancient inability to fold their wings flat over the body. They differ from all other insects in their combination of biting mouthparts; their two equal (or almost equal) pairs of long,

Introducing Dragonflies

Representing the two major groups of dragonflies: the slim Western Red Damsel (left, a female) is a damselfly and the more robust Four-spotted Skimmer (right, a male) is a true dragonfly.

membranous, net-veined wings; their large, bulging eyes and short, thread-like antennae; and their long, slender abdomen that, in the male, bears secondary genitalia at the base.

A dragonfly leads a dual life – in its immature stage, the larva lives in water, obscure and camouflaged. When it is time to mature, the changing larva emerges from the water and transforms into a colourful, flying adult.

Some early dragonfly-like insects were enormous – fossils from the Carboniferous Period show that one had a wingspan of 70 cm – but the largest North American species found today measures about 14 cm across the wings. The greatest wingspans in modern times – about 17 cm – belong to the giant damselflies of the American tropics.

Many dragonflies around the world are as colourful and flashy as the most spectacular birds and butterflies. Most of our local dragonflies are more subdued, but they are still lovely and striking insects. They come in a myriad of colours, from iridescent metallic green to breathtaking crimson. Their bodies can be boldly spotted or striped, and their wings are often strongly patterned with spots and bands of colour.

The order Odonata is divided into three suborders: the Zygoptera (damselflies), the Anisoptera (true dragonflies) and the Anisozygoptera (a tiny group of two rare species from the mountains of eastern Asia). Damselflies are slimmer, often smaller and usually fly more slowly than true dragonflies. At rest they usually hold their equal-sized wings together above the body – Zygoptera means "joined wings". Anisoptera means "unequal wings", because the hindwings of the true dragonflies are broader than the forewings. When perched they hold their wings out away from the body.

Introducing Dragonflies

A male Paddle-tailed Darner: performance and efficiency in flight.

The flying ability of dragonflies amazes most people. Although the wing structure and arrangement of the flight muscles are primitive, the flight performance and efficiency are remarkable. Unlike most insects, dragonflies usually beat their forewings and hindwings separately – when the forewings are up, the hindwings are down. Each wing also has much independent control, accounting for the surprising manoeuvrability of many species, which can fly upwards, sideways, backwards and forwards. A large darner can fly up to 60 km per hour. Darners, emeralds, spiketails, river cruisers and some skimmers are called *flyers* because they spend most of their active life flying – they even generate additional body heat from their wing muscles. Damselflies, clubtails and most skimmers are often called *perchers*, because they spend more time perching than flying. Perchers gain much of their body heat from basking in the sun and make only short flights to catch food or mate.

For millennia, dragonflies have instilled superstitious fear in humans, even though they do not sting or bite people. Maybe their boldness takes us aback, or their speed startles us. To the uninitiated, their strange appearance up close can make them seem fearsome. The English name "dragonfly" echoes the feelings these insects sometimes arouse – they are the fanciful "devil's darning needles" that sting venomously or sew up the lips; they are "snake doctors" with the power to bring dead snakes back to life. These legends and folktales are groundless – dragonflies are harmless to humans.

Large and abundant, dragonflies form one of the predominant groups of invertebrates in freshwater communities in British Columbia and the Yukon. They live around most types of fresh water but, in our western mountains and valleys, they are less abundant in running water than they are in standing-water habitats. Some kinds prefer lake shores, others are found only along streams, around springs or in peatlands. Ponds and marshes rich in aquatic vegetation support the most species.

Larvae are strikingly different from adults. The Pacific Forktail larva (a damselfly) is slender and has three leaf-like gills, while the Four-Spotted Skimmer larva (a true dragonfly), is stocky and its gills are internal.

Life History and Behaviour

Dragonflies spend their youth as aquatic larvae preying on other underwater animals. Dragonfly larvae – sometimes called nymphs – have an enormous (for their size) hinged labium, a sort of lower lip armed with pincers, that they use as an extendible grasping organ for capturing prey. They are voracious predators, eating small aquatic insects, crustaceans and even fish and tadpoles.

Biologists place dragonfly larvae into three categories, according to their feeding behaviour: *Claspers* (damselflies and darners) stalk their prey while using their clasping legs to hold onto vegetation; their colour patterns of green and brown help camouflage them among the water plants. *Sprawlers* (river cruisers, emeralds and most skimmers) lie spread-eagled on the bottom mud and debris or on vegetation, waiting to ambush prey; they often keep hidden under a coating of mud and algae. *Burrowers* (clubtails and spiketails) dig into the sand and silt, where they await their prey.

Damselfly larvae, like the adults, are slender animals. The tip of the abdomen bears three leaf-like gills, richly laced with the fine tubes that carry oxygen and carbon dioxide throughout the body. The stouter larvae of the true dragonflies do not have external gills; instead, they pump water in and out of the gut and breathe through gills lining the rectum. Damselflies use their gills to help them get around, sweeping them back and forth like swimming fins. Larvae of true dragonflies also use their breathing mechanism to help them move: they can blast pressurized water out the anus, jet-propelling them through the water – an effective tactic for escaping predators or attacking prey.

Dragonflies, like grasshoppers and many other insects, develop without a pupal stage. After the larva pops out of the egg, it eats, grows and moults 8 to 17 times (usually 10 to 14), depending on the species and the conditions. The developing wingbuds get larger with each moult. For

Introducing Dragonflies

Left to right: a male Variable Darner emerging as an adult.

many species in British Columbia, the life cycle takes about a year. Some spreadwings and meadowhawks that live in temporary ponds overwinter as eggs, hatch in the spring, grow rapidly and emerge as adults in the summer. Many species overwinter as larvae and emerge the following spring or summer; others spend two years in the larval stage. For some dragonflies (especially certain darners and emeralds), the larval life may last six years or longer. Development time depends on the species and also on altitude, latitude and amount of daylight. Growth slows with the shorter summers and colder temperatures of northern habitats and high altitudes.

In British Columbia and the Yukon, dragonflies live only a short time as adults – about one to two months. A dragonfly begins its adulthood when the fully grown larva metamorphoses into an adult inside its last larval skin, then crawls out of the water, up a plant stalk or some other support. Clubtails and pond damsels can emerge horizontally on rocks, floating logs and plants, or the shore.

Now exposed to air, the dragonfly begins its final moult: the top of the thorax splits open and the adult dragonfly squeezes out of the larval skin. It pumps blood into its wings and abdomen, which expand slowly, and gradually, the body hardens. After an hour or two the dragonfly can fly, weakly at first, on fragile, glistening wings. It leaves the empty larval skin, the exuvia, clinging to the support. Once its body has hardened, the adult dragonfly will not grow larger even though it eats a lot.

Emergence can occur by day or night. Most damselflies, clubtails and some skimmers emerge during the day. Many darners emerge at night, but in cool weather or in the far north, may transform during the day. The newly emerged adult, called a teneral, is vulnerable to predators and bad weather.

Life History and Behaviour

A teneral female Red-veined Meadowhawk with her exuvia.

Dragonflies have characteristic flight periods. This is the period during which adults may be seen and does not necessarily represent the adult lifespan of a particular individual. Many species may live in the same locality: some emerge early in the spring and are rare by summer; others appear in mid summer and fly into the fall; still others fly from spring to fall.

After emergence, most adults leave the shoreline to hunt and eat for a few days or even weeks as they mature. They are powerful predators that hunt by sight. They usually capture prey while flying, grabbing it with long, spiny legs and then chewing it with powerful jaws. Adult dragonflies eat mainly flying insects, but some species will pluck insects, spiders and even small frogs off vegetation or the ground.

Immature adults are pale in colour, but gradually become darker and often brighter as they mature. Some species produce a waxy, white or pale blue powder, called pruinescence, over parts of the body and wings; this is especially obvious in some male skimmers and spreadwings and some female forktails.

When they are sexually mature, dragonflies return to the water to breed. Most of the dragonflies you see near water are males aggressively searching for mates. In many species (damselflies, petaltails, clubtails, emeralds and many skimmers), mature males defend a territory against other males of the species, patrolling the habitat or sally-

A male Canada Darner feeding on a meadowhawk.

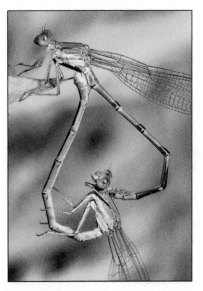 

Above left, a couple of Emma's Dancers (damselflies) form an upside-down heart; and above right, two Variable Darners mate on a twig.

ing out from perches. This territorial behaviour limits aggression by spacing males along the shore and helps prevent undue disturbance of egg-laying females.

Females coming to the water to breed quickly attract mates. With the appendages at the tip of the abdomen, a male grasps a female by the front of the thorax (damselflies) or by the top of the head (true dragonflies). This head to tail arrangement is called the tandem position. Before joining with a female or even while in the tandem position, the male transfers sperm from the tip of his abdomen to his penis, which is under the second abdominal segment. The female then loops the end of her abdomen up to the penis so that the male can transfer the sperm to her. The Odonata are the only insects that mate in this circular formation, called the wheel position, which they maintain for a few seconds or several hours, depending on the species. Female dragonflies usually mate more than once, and in an attempt to ensure that their sperm fertilizes her eggs, males may spend much of the copulation period removing the sperm of other males – the penis is modified to pull another's sperm out of the female or push it aside so that it is inactivated.

The female lays her fertilized eggs by the hundreds. All damselflies, darners and petaltails have a knifelike egg-laying structure with pointed blades, called an ovipositor, at the tip of the abdomen; they lay their eggs in plant tissue, although some darners and petaltails insert eggs into soil.

Life History and Behaviour

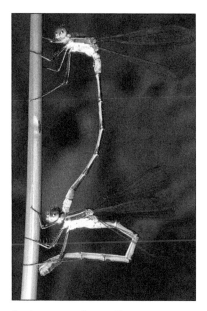

Laying eggs: above, Common Spreadwings, in tandem, deposit eggs in the stalk of a plant; above right, a lone Paddle-tailed Darner places eggs in a wet log; right, a pair of Cardinal Meadowhawks drop eggs in water.

Spiketails shovel the eggs into a streambed. Other species – without ovipositors, or with simply a scoop-shaped plate called a vulvar lamina – drop eggs into the water, tap the eggs into mud and moss, or simply dip the tip of the abdomen into the water and wash the eggs off.

Competition for mates is usually fierce, and male aggression can prevent females from laying their eggs. Females that lay their eggs alone, especially mosaic darners, often do so stealthily, flying low among the plants along the shore, wings rustling in the stems as they settle to deposit the eggs. Some damselflies actually crawl below the water's surface to escape the attention of males, often remaining submerged for more than an hour – they take a film of air down with them, trapped in the hairs on their body, so they can breathe while they lay their eggs. In many damselflies, the meadowhawks and the Common Green Darner, the male stays in tandem, retaining his hold on the female while she lays her eggs. In some other species, the male hovers protectively nearby, guarding the egg-laying female from any other males who may attempt to mate with her, and allowing her to lay her eggs undisturbed.

Dragonfly Habitats in British Columbia and the Yukon

Dragonflies live in many freshwater habitats and, with practice, you can predict the species you will see when visiting different types of water bodies. Below are the general types of habitats that dragonflies prefer in B.C. and the Yukon. The species mentioned for each habitat typically breed and develop in these places. Of course, what you see will depend on where you are and when – many species live only in certain regions and fly at certain times. Places with a mixture of habitat types usually have the greatest diversity of species.

Large lakes. The shores of large lakes, wave-washed and with little aquatic vegetation, support few dragonfly species. But in warm southern valleys, several species can live in this habitat: Marsh and Tule bluets (in bulrush beds), Emma's Dancer, Shadow Darner, Pale Snaketail, Pronghorn Clubtail and Western River Cruiser. Some of these species are also typical of warm rivers. The large, deep lakes north of the southern valleys are usually too cold and unproductive; there, dragonflies are restricted to shallow waters in sheltered bays, where the species are usually the same as those found in small lakes or ponds.

Small wooded lakes and ponds. These water bodies frequently support floating plants such as waterlilies, but have little emergent vegetation; they often have peaty edges. A number of species are typical of small lakes and ponds in wooded areas: Hagen's and Marsh bluets; Black-tipped, Blue-eyed, Canada, Lake, Paddle-tailed and Shadow darners; Beaverpond and Spiny baskettails; American, Lake and Ringed emeralds; Crimson-ringed, Hudsonian and Red-waisted whitefaces; Chalk-fronted Corporal and Four-spotted Skimmer; Yellow-legged Meadowhawk.

An exposed shore of Osoyoos Lake in B.C.'s southern interior.

Portal Lake in Mount Robson Provincial Park is a small wooded lake with waterlilies.

Habitats

Marshes on the Columbia River, near Invermere, feature cattails and bulrushes.

A sedge marsh at Horsethief Creek, near Invermere in southeastern B.C.

Cosens Bay Pond (Kalmalka Lake Povincial Park, near Vernon) dries up in the summer.

Cattail/bulrush marshes (including the marshy margins of lakes and ponds). These still-water habitats are mostly in warm lowland areas. Dense beds of cattails (*Typha*), with their sword-shaped leaves and cylindrical brown flower heads, and bulrushes or tules (*Scirpus*), with tall, tubular, leafless stems, are the characteristic plants in these places. These marshes are rich in plant and animal life. Common, Emerald, Lyre-tipped, Spotted and Sweetflag spreadwings; Marsh, Northern and Tule bluets; Pacific and Western forktails; Blue-eyed, California, Canada, Lance-tipped and Variable darners; Common Green Darner; Eight-spotted, Four-spotted and Twelve-spotted skimmers; Common Whitetail; Blue Dasher; Dot-tailed Whiteface; Western Pondhawk; Black, Cardinal, Cherry-faced, Red-veined, Saffron-winged, Striped, Western and White-faced meadowhawks.

Sedge marshes. Sedges (*Carex*) are common, grass-like plants that grow up to 150 cm tall in dense stands around lakes and ponds or in large areas of shallow water. Common, Emerald, Spotted and Sweetflag spreadwings; Boreal, Northern and Taiga bluets; Sedge Sprite; Paddle-tailed, Sedge and Variable darners; Hudsonian and Mountain emeralds; Four-spotted Skimmer; Boreal and Hudsonian whitefaces; Cherry-faced and White-faced meadowhawks.

Temporary ponds. Any ponds that dry up in summer. Emerald, Lyre-tipped and Spotted spreadwings; Cherry-faced, Red-veined, Striped and Variegated meadowhawks. Some of these species overwinter as eggs in the dry pond basin.

Alkaline lakes and ponds. These habitats occur primarily in the dry, warm southern valleys and plateaus. No species is restricted to these unusual habitats, but some are able to live in them despite the water's high salt content, and their life histories enable them to take advantage of the temporary nature of the shallower lakes and ponds. Lyre-tipped and Spotted spreadwings; Alkali and Boreal bluets; Saffron-winged and Variegated meadowhawks.

Peatlands. Peatlands are aquatic or semi-aquatic habitats where plant decomposition is so slow that peat accumulates. Bogs are acidic, low-nutrient peatlands whose water comes only from rain and snow; they are dominated by sphagnum mosses. Fens are peatlands affected by flowing ground water and, thus, richer in minerals and less acidic than bogs; they are dominated by sedges, grasses and non-sphagnum mosses. Peatlands are complex habitats, but there are two distinct dragonfly habitats. (1) Where aquatic mosses float in fens and bog ponds, several species are typical around the water's edge: Subarctic Bluet; Azure and Subarctic darners; Muskeg and Treeline emeralds. (2) Many fens have large mats of moss, often covered with shallow water or with scattered shallow pools and puddles (often drying in summer), and supporting sedges in various densities: Common and Sweetflag spreadwings; Boreal and Taiga bluets; Sedge Sprite; Black-tipped and Zigzag darners; Delicate, Mountain, Quebec and Whitehouse's emeralds; Black Meadowhawk; Canada and Hudsonian whitefaces.

Alkaline pond at Cherry Creek, near Kamloops.

Peatland habitats: Centre, a pond with aquatic moss in the Porcupine River drainage, Yukon. Right, a mossy fen with shallow puddles near Takla Landing, B.C.

Habitats

Tezzeron Creek, near Fort St James, is a relatively warm stream flowing from a lake.

Net ready, I search for Forcipate Emeralds at a streamlet in a subalpine fen in the Rocky Mountains.

Streams. Dragonflies are not normally found in the cold mountain streams of British Columbia and the Yukon. Species living in flowing water are generally restricted to warm, southern lowland streams or mountain streams that drain warm lakes, beaver ponds or peatlands. River Jewelwing, Emma's Dancer, Sinuous Snaketail, Olive Clubtail, Pacific Spiketail and Western River Cruiser are restricted to southern valleys and plateaus; Shadow Darner and Pale Snaketail are more widespread and also live in lakes; Brush-tipped and Ocellated emeralds live along small montane streams throughout much of British Columbia; Boreal Snaketail has been found only in warm creeks and rivers north of about 52°N.

Springs. Some of the more uncommon dragonflies of the region live in small springs and the shallow trickles and pools associated with them. The Vivid Dancer is mostly restricted to small streams that drain hot springs in the mountains, although it occurs in some small, spring-fed streams in the southern valleys. The Pacific Spiketail is found in many warm streams draining lakes on the west side of the Coast Mountains, but in the interior, it occurs rarely and mostly along spring-fed streams. The Black Petaltail lives in mossy wet places in the Coast Mountains where the larvae burrow in the mud. The Forcipate Emerald breeds in spring-fed trickles in the Rockies and northern mountains. More common, the Western Red Damsel is widespread in shallow ponds in southern British Columbia; these ponds are often, but not always, associated with springs. The only records of the Plains Forktail in our region are at Liard River Hot Springs.

Studying Dragonflies

Over many years of study, biologists have compiled a significant amount of information about dragonflies, but we still have a lot to learn. The geographical distribution of many species is poorly known. We need more thorough descriptions of the behaviour of many species in our region, as well as carefully documented details about adult and larval habitat preferences and long-term observations of species at a particular site. As with the study of other organisms, dragonfly experts welcome input from laypersons interested enough to collect data on these insects.

If you are interested in studying dragonflies, a good way to start is by taking pictures. Dragonflies are particularly photogenic insects, and photography is a satisfying and useful way to study them. Pictures showing behaviours such as emergence, feeding, mating and egg-laying are especially valuable. Dragonflies are often hard to get close to, but patience pays off, and you'll be able to get good photographs of many kinds, especially those that perch frequently. A single-lens reflex camera with a close-focusing (macro) lens is best. Lenses with a long focal length for increased magnification work well; use screw-on close-up lenses or extension tubes for close focusing. An electronic flash will improve the depth of field and reduce the effects of movement.

Dragonflies that fly more than perch (darners, emeralds and others) are difficult to photograph. Some photographers take a cooler into the field; they place a captured dragonfly in an envelope or plastic bag and cool it for a while in the cooler. Then they hang the dragonfly in a natural pose on a leaf or twig (these groups of dragonflies usually perch by hanging) and take the picture. This procedure requires patience – the dragonfly may warm up and fly away before you can focus and shoot.

Most observers are interested only in watching dragonflies. Some will want to photograph them or catch, identify and release them. But the seriously interested student might want to collect some specimens. Collections provide information that is impossible to get from live insects in the field, and some species cannot be identified with certainty without microscopic examination. Even a small collection made in an area where few dragonflies have been recorded will improve our understanding of dragonfly distribution and habits.

Collecting a few dragonflies will not harm a population, but collections should be made only for particular research or educational purposes. Larvae and exuviae can also be collected and offer just as much information as adults do, although similar species are sometimes difficult to distinguish. Collection information (exact location, date, habitat, collector) should always be kept with the specimens, which should be stored in airtight containers. All collections should be available for study by other researchers.

Collections of dragonflies at the Royal British Columbia Museum in Victoria and the Spencer Entomological Museum in the Zoology Department at the University of British Columbia are important resources for the study of dragonfly identification, evolution and biology. See p. 93 for a list of books and web sites containing detailed information on dragonflies and collecting.

Dragonfly Conservation

The most serious stress on dragonfly populations is the elimination or alteration of their freshwater habitats. The most destructive practice is the draining and filling of marshes and ponds. Dam building is also harmful, flooding many important habitats and creating few in return. The destruction of natural lakeshores when building houses and swimming beaches is reducing the habitat of many lake dwellers, particularly in the warm southern valleys. Cattle often trample and pollute small ponds or spring-fed streams in grasslands and dry forests where several dragonfly species breed. Logging and associated road building can cause problems in streams – erratic flow, warmer water temperatures and higher silt content – which can harm dragonfly larvae. Logging can also harm natural communities in peatlands, marshes and lakes, especially at higher elevations.

Global climate change will affect the distribution of dragonfly habitats in our region. There will be less water available in most areas (especially in summer) as temperatures and evaporation increase. Shallow lakes and ponds, particularly in warmer areas, may disappear. Groundwater will play a more important role than it does today, and many water bodies will become more alkaline. The additional nutrients will change water chemistry. Peatlands will become more marshy and our fascinating fauna of bogs and fens may decline.

What can we do? We must learn more about our dragonflies so that we can protect them and their habitats. But also we must act quickly to protect the natural communities where dragonflies live. Small ponds, marshes, springs and streams in British Columbia's southern valleys sustain some of our rarest species, and these habitats are among the first to disappear when we expand housing, industrial and agricultural developments. Even in remote areas sensitive aquatic ecosystems can be drastically affected by industrial activity. Suitable dragonfly habitats are disappearing faster than new ones are being formed, so there is increasing cause for concern.

We can get involved with local naturalist organizations to learn more about the natural world in our neighbourhoods. We can encourage all levels of government to protect aquatic ecosystems on public land. We can get involved in public processes to develop land-use plans and regulations that preserve, rather than destroy, natural diversity. And we can maintain and create natural habitats on our own property. Build a pond in your garden or on your farm (ponds without fish are best) and see what arrives. Learn about the dragonflies and other creatures that settle there and encourage others to do the same.

How to Find and Identify Dragonflies

Dragonflies can show up almost anywhere – some species fly far away from the place where they grew up. But, by far, the best way to study them is to visit a pond or shallow lake, rich with aquatic vegetation – go in the middle of a sunny day in July or August and you'll easily see 5 or 10 species, maybe even more. Watch them, figure out what they are doing and try to identify the species. Soon you'll be able to recognize the different families and the more common species, and distinguish males and females. If you frequent your favourite local pond, you'll see that different species fly at different times of the year. Try to visit other kinds of habitats, because many species have particular preferences for certain kinds of ponds or streams. Go even farther afield, because different species live at higher or lower elevations, or in different parts of British Columbia and the Yukon.

Maybe all you want to do is watch dragonflies and try to name the most distinctive ones by comparing them to the pictures in this book. But if you are keen to identify as many kinds as you can, you'll have to delve further into the species descriptions. Whatever your level of interest, here is how you can identify dragonflies once you have found them:

Step 1: Look carefully at the size, shape and colour of the dragonflies. Concentrate on a particular individual. How does it fly? How does it perch? How does it hold its wings? Are the eyes widely separated? What colours and patterns do you see? Some species are very similar to others, so small details are important. Binoculars are useful for field identification; pick a pair with a magnification around 6x to 8x that focuses closer than eight feet.

Step 2: The best way to identify a dragonfly is to catch it in an insect net and identify it in the hand. Gently hold the wings together between finger and thumb, with the wings over its back; this will not harm the insect. But *do not hold a teneral dragonfly* – you will damage its wings. You

may have to use a 10x hand lens to see some features well. Look carefully at the dragonfly and release it after you have made your identification.
Step 3: Use the Family Key (page 23) to narrow your choices. Before long, you'll be able to recognize the families without the key.
Step 4: Compare your dragonfly with the photographs in this book, checking for similar shape, colours and patterns. The photographs may narrow your selection to the genus or, for some, the species.
Step 5: Check the characteristics in the text to find the species or confirm your identification from the photograph.

Consider all the information: appearance, behaviour, geographical location, habitat and flight period. With this book, you should be able to identify most species. Some are easy to recognize at a distance. A few are difficult, or perhaps impossible, to identify in the field, even after close inspection – especially some females and some tenerals that do not have their mature colours.

The Dragonfly Body

Dragonflies, like other insects, have three major body divisions: head, thorax and abdomen (fig. 1, next page). The head is large and mobile, connected to the thorax by a slender neck. The eyes are large and prominent, often occupying most of the head. The eyes of skimmers, emeralds and river cruisers meet broadly on top of the head; a spiketail's eyes barely touch; those of clubtails and petaltails are separated; and damselfly eyes are smaller and farther apart than those of clubtails and petaltails. A dragonfly eye consists of many small, single-lens eyes that work together to form an image; some large species may have 30,000 lenses in each compound eye. Their vision is excellent and they are specially good at detecting motion. Three single-lens eyes (ocelli) between or in front of the compound eyes may help regulate stability in flight. The antennae are short, hair-like and, among other things, probably give the dragonfly a sense of how fast it is flying. Dragonflies apparently do not sense sound. The powerful mouthparts, dominated by the chewing mandibles, lie under the head.

The thorax is the dragonfly's locomotive centre. The small, mobile prothorax carries the front legs. The middle and hind legs and the two pairs of wings are attached to the large, fused, second and third segments of the thorax, which are mostly filled with flight muscles. The wings are densely veined and the vein pattern is sometimes important in identification. The presence and shape of the anal loop, a cluster of cells in the hindwing, is often a useful clue to family identification. The pterostigma, the coloured cell on the front of the wing near the tip, helps the wing twist in flight, an important aspect of flight dynamics. It may also serve as sort of a flag or signal in sexual interactions.

The long abdomen stabilizes flight and contains the mating structures. It can also act as a radiator, dissipating the heat generated by flying. The abdomen has ten segments, which we refer to by number from the front;

Finding and Identifying Dragonflies

Figure 1. Lance-tipped Darners, male above and female below, showing some body parts important in identification.

but it is usually easier to count back from segment 10 at the tip. The male genitalia, located underneath segments 2 and 3, consist of complex structures, including the penis and the hook-like hamules, which hold the female's abdomen during mating. Female damselflies, darners and petaltails have a knife-like ovipositor developed from segments 8 and 9. The spiketail's ovipositor is like a narrow shovel; in the clubtails, river cruisers, emeralds and skimmers, it is reduced to a scoop-shaped or scale-like vulvar lamina (or subgenital plate), which holds the eggs while they are laid. Both males and females have appendages on segment 10 at the tip of the abdomen that can be important in identification. Males have a pair of upper appendages; damselfly males have two lower appendages, while other dragonfly males have a single lower one. Females have a pair of upper appendages, usually small and inconspicuous, but large and leaf-like in darners; they have no obvious lower appendages. The male uses its appendages for clasping the female during mating. In any species, the female's abdomen is thicker than the male's; it contains the ovaries and eggs.

Key to Dragonfly Families in Our Region

This key will help you identify families of B.C. and Yukon dragonflies in the field. Keys are short-cuts to identification. The paired statements – "a" and "b" for each number – contain contrasting characteristics. Look at the dragonfly and compare it to the two statements: pick the statement that agrees with the dragonfly's characteristics – then go to the next pair of statements indicated by the number following. Keep going until you reach a family name. To use this key successfully, you will need to see the dragonfly up close or have it in your hand. Once you become familiar with the families, you will no longer need the key.

1a. Wide head with eyes separated by more than their diameter; forewings and hindwings similar in shape, the hindwing not broader at base than the forewing; wings held together over the abdomen (in one family mostly open, but not flat, at about 45°); slender bodied and mostly small. . . . . Damselflies (Zygoptera) – 2
1b. Eyes closer together than their diameter (touching in most species); hindwings much broader at the base than the forewings; wings held wide open, flat or angled downwards; heavier bodied and mostly medium-sized to large.
. . . . True Dragonflies (Anisoptera) – 4
2a. Wide black tips on the wings; length, 45 mm or more.. Jewelwings
2b. Clear wings (except for the small rectangular pterostigma on the front edge near the tip); length, 40 mm or less. 3
3a. Long pterostigma, equal to the diameter of an eye; wings held open at about a 45° angle (but may be closed in tenerals or when the specimen is handled in a net). . . . . Spreadwings

3b. Pterostigma shorter than the diameter of an eye; wings closed over or above the abdomen (may be partially open when mating or when handled in a net). ....Pond Damsels
4a. Eyes completely separated. 5
4b. Eyes just touching or meeting broadly on top of the head. 6
5a. Black thorax with numerous yellow spots. .... Petaltails
5b. Thorax not coloured as above.........................Clubtails
6a. Longer than 70 mm; eyes barely touching at a single point; black body with two yellow stripes on each side of the thorax and yellow spots on top of the abdomen. .... Spiketails
6b. Characters not as above; if large with pale stripes on thorax sides, then the eyes meet broadly on top of the head. 7
7a. Longer than 55 mm; abdomen spotted or otherwise marked with a paler colour..8
7b. Almost always shorter than 55 mm; abdomen almost all black in species longer than 50 mm (Lake Emerald is usually longer than 55 mm and has thin white rings on a black abdomen.)...........9
8a. Each abdominal segment has a single yellow spot on top (sometimes a small pair of spots on segments 3 to 6); thorax has a single pale stripe on each side. .... Cruisers
8b. Each abdominal segment has multiple spots, or none; thorax has two stripes on each side, or none...................... Darners
9a. Abdomen almost all black (sometimes with thin white rings); eyes bright green or turquoise............................. Emeralds
9b. Variously coloured abdomen; if all black, then the eyes are not green; if eyes are green, then the body is not black..............10
10a. Hindwing's anal loop foot-shaped with a short, blunt "toe" (fig. 2b). .... Emeralds
10b. Anal loop has a longer, pointed "toe" (fig. 2c)...........Skimmers

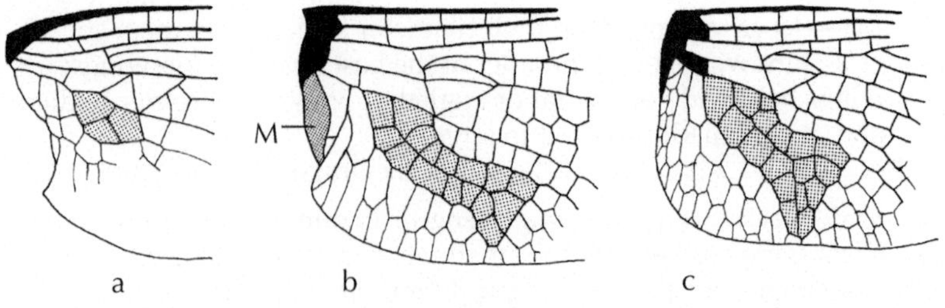

Figure 2. Anal loops: a, snaketails (represented by the Pale Snaketail); b, emeralds (American Emerald); c, skimmers (Yellow-legged Meadowhawk). M = membranule.

The Dragonflies of British Columbia and the Yukon

In the following pages I briefly describe the 88 species of dragonflies that we know live in our region: 87 in British Columbia (B.C.) and 33 in the Yukon. I also outline each family and genus, but omit the genus description if there is only one species in the family. Families are listed in phylogenetic order, which reflects our understanding of the sequence of dragonfly evolution. The genera and species are in alphabetical order of their English names.

The species descriptions list most of the key characteristics, as outlined in The Dragonfly Body. The statements in the following descriptions refer to adults of both sexes unless otherwise indicated. To keep this book functional for most people, I have kept technical details to a minimum, which will make some identifications very difficult. Those who wish to make the more difficult identifications should consult a more scientific source, such as the ones listed on page 93. The length is the average body length in millimetres for males (♂) and females (♀) taken from specimens in our region.

Range tells the worldwide and regional distribution. Most species described in this book are restricted to North America, and for brevity, I use one-word terms to indicate the species ranges there. These terms are defined in the box on the next page.

Field notes contains other relevant information about the species, such as its rarity, preferred habitats, behaviour and flight period. The flight period is my estimate of the length of time adults can be seen throughout B.C. or the Yukon, and is often longer than the flight period in any one location. This estimate is based on museum specimen records, reliable sight records, records from nearby states and provinces, and extrapolations from these records and other biological clues.

Range Types

Each species lives across a particular geographical range. For brevity in the Range section of the accounts, I have lumped many similar North American dragonfly distributions into range types, as described below. The ranges of a few species extend beyond this continent, and I have indicated these other places after the range type.

East Beringian: Confined to extreme northwestern North America, presumably having lived in the Beringian glacial refugium (a huge ice-free area centred on the Bering land-bridge) during the last glacial period. The only species here is the Treeline Emerald.

Montane: Living in the western North American mountains and the associated plateaus and valleys, including the coastal lowlands.

Northern: Occurring in the northern spruce forests, across the continent's boreal zone. In general, these species range from the Atlantic provinces, across the northern New England states, Quebec, northern Ontario, parts of the northern tier of Midwestern states, the Prairie provinces north of the Great Plains, to northern British Columbia, often ranging considerably southward in the higher mountains and plateaus.

Pacific Coastal: Confined to the lowlands of the Pacific coast.

Southern: Transcontinental in southern North America and, at the northern extremities of their ranges, entering Canada to varying degrees along the border.

Transition: Generally most common in the southern boreal forests and adjacent dry forests in the West and the mixed forests and deciduous forests in the East.

Western: Ranging widely west of the 100th meridian in North America, but not at all to the east.

Widespread: Broadly distributed over all of North America, overlapping several of the other range types. Some of these species range into boreal regions.

Facing page: Regional map, showing some major features.
(Produced by Clover Point Cartographics, Victoria, from a North American base map, 1995, of Natural Resources Canada, Ottawa.)

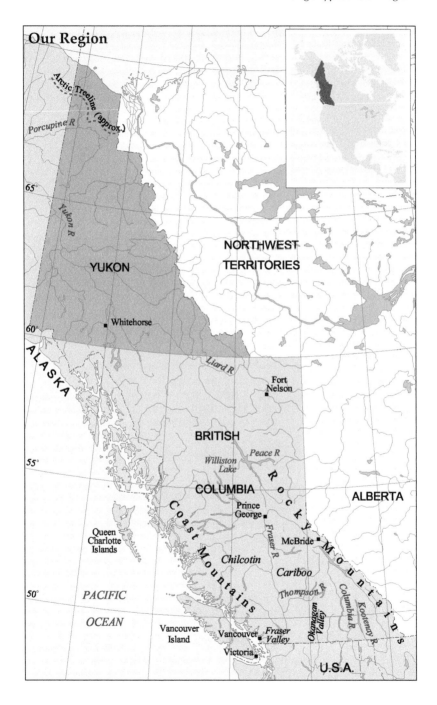

DAMSELFLIES   Suborder Zygoptera

Jewelwings   Family Calopterygidae

The most spectacular damselflies in Canada: large with metallic-coloured bodies and broad, densely veined wings, often coloured. Larvae develop in streams, clinging to plant stems and roots; they can be distinguished by their long antennae – the first segment is longer than all the others combined. Only one jewelwing lives in British Columbia.

River Jewelwing
Calopteryx aequabilis
Metallic green; males have blue reflections. Long, black legs bear long spines. Males have dark-brown wingtips and no pterostigmas. Females' wing bases are shaded yellow-brown; wing tips are brown, pterostigmas white. Length: ♂ ♀ 47 mm.

Range: Transition. In B.C., found only along Christina Creek at Christina Lake in the southern interior.

Field notes: Lives along small warm rivers and slow streams where the adults fly with a dancing, butterfly-like motion and perform striking courtship displays. Flight period: B.C., mid June to early September.

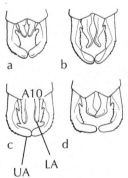

Figure 3. Spreadwings, male appendages, top view: a, Spotted; b, Lyre-tipped; c, Common; d, Emerald.
A10 = abdominal segment 10;
UA, LA = upper, lower appendages.

Figure 4. Spreadwings, tips of female abdomen: a, Common; b, Sweetflag. A9 = abdominal segment 9; OV = ovipositor.

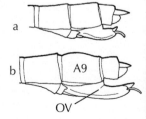

Spreadwings   Family Lestidae
A small but widely distributed family in B.C., containing only one genus here, *Lestes*, with five species. The common name comes from the characteristic posture of the adults – they usually perch with wings half-spread.

Spreadwings   *Lestes*
Large damselflies, brown, black, metallic-green or bronze above, mostly pale below. As they age, parts of the body, including the tip of the abdomen in males, often become pruinose bluish white. Females lay eggs in tandem with males, usually in plants above the surface of the water. The larvae are long and slender with banded gills and an unusually elongated labium. Some species are adapted to temporary ponds; the eggs overwinter and the larvae grow rapidly after the basin fills with water in the winter or spring.

Spotted Spreadwings laying eggs: ♂ above, ♀ below.

Spotted Spreadwing   *Lestes congener*
Thorax top is dark brown or black with narrow pale stripes; the pale underside has several (usually three) dark spots on each side. Male's appendages, fig. 3a. Length: ♂ ♀ 40 mm.

Range: Widespread. And widespread in B.C., especially in the south.

Field notes: Common in many types of wetlands, from alkaline ponds to cattail marshes, lakes and peatlands. In B.C., it emerges later than other spreadwings and is the last damselfly seen in autumn. Flight period: B.C., early June to mid November (flight before early July is unusual).

Lyre-tipped Spreadwing *Lestes unguiculatus*

Similar to the Common Spreadwing, but the tips of the male's lower appendages curve outward like the shape of a lyre (fig. 3b), which gives the species its English name. On most, the back of the head is mostly pale, especially in females, which do not darken with age as much as males. Length: ♂ 37 mm, ♀ 35 mm.

Range: Widespread. In B.C., found south of about 56°N; distinctly more common in the southern valleys.

Field notes: Typical of warm ponds at low elevations, especially temporary and alkaline ponds. Less commonly, it lives in peatland pools and sedge fens in the mountains. Flight period: B.C., early June to mid September.

Emerald Spreadwing *Lestes dryas*

The thorax top is metallic green. Male's appendages, fig. 3d. The end of the female's ovipositor reaches the tip of the abdomen.
Length: ♂ 36 mm, ♀ 35 mm.

Range: Widespread; also across northern Eurasia. Throughout most of B.C., and in Yukon valleys north to the Porcupine River drainage.

Field notes: Usually the first spreadwing to emerge in the year. In the south, it is common in early summer, especially around small ponds and places that may dry up in summer. It is less common northwards and uncommon in the Yukon. Flight period: B.C., late May to mid September; Yukon, early July to late August.

Common Spreadwing *Lestes disjunctus*

The mature male's thorax has blue stripes on top, but turns pruinose grey with age; eyes are dark above and pale blue below. On both sexes, the back of the head is black; this is especially useful in separating female Common and Lyre-tipped spreadwings. Male's appendages, fig. 3c. The end of the female's ovipositor does not reach the tip of the abdomen (fig. 4a).
Length: ♂ 37 mm, ♀ 35 mm.

Range: Widespread. The most common spreadwing in our region, north to the Arctic treeline.

Field notes: Common around most types of standing water with abundant aquatic vegetation. In most places, Common Spreadwing adults emerge after the first Emerald Spreadwings but before most Spotted Spreadwings. Flight period: B.C., mid June to mid October; Yukon, late June to mid September.

Sweetflag Spreadwing *Lestes forcipatus*

Male very similar to the Common Spreadwing. Identification is best done with the female, whose large, distinctive ovipositor extends beyond the end of the abdomen (fig. 4b). Length: ♂ ♀ 40 mm.

Range: Southern. Scattered localities in B.C. south of 56°N.

Field notes: Unknown in B.C. until 1998; it is so similar to the Common Spreadwing, it had been overlooked. Since then, although it is uncommon, it has been found in a variety of ponds, marshy lakes and peatlands, but is probably most common in sedge fens. In eastern North America, at least, this damselfly commonly lays eggs in the sweetflag, a kind of aquatic iris, thus its English name. Flight period: B.C., mid June to mid September.

Pond Damsels   Family Coenagrionidae
Small damselflies that normally perch with wings closed above the abdomen. Most males are blue marked with black, but the main colour may be green, yellow, orange, red or purple. Females often have two colour forms per species, one similar to the male (usually blue). Females lay eggs in the tissues of water plants, sometimes completely submerging themselves for a long time while laying. Larvae are not as long as spreadwing larvae and have short labia, unstalked at the base. There are six genera and 18 species of pond damsels in our region. The American Bluets (*Enallagma*) and forktails (*Ischnura*) are the most common groups.

Dancers   *Argia*
The largest pond damsels in B.C. Males are blue or violet, with black markings; females are the same, or olive or brown. Dancers usually develop in streams. Adults like to rest on bare sunny spots by the water. Larvae are stocky and squat, their gills unusually broad and pigmented. Adult dancers can be distinguished from bluets by isolated black marks on the sides of the abdominal segments and by a short black stripe on the side of the thorax, which narrows in the middle. The female has no vulvar spine in front of the ovipositor.

Vivid Dancer   *Argia vivida*
Male is bright blue; female is brown or blue. The top of the thorax has a black stripe at least as wide as adjacent pale stripes.
Length: ♂ 34 mm, ♀ 35 mm.

Range: Montane. Southern B.C. from the Coast Mountains to the Rockies.

Field notes: Rare. Associated with cool or hot springs. The larvae live in the pools and outflow streams that are smaller and more vegetated than those preferred by Emma's Dancer. The small populations of Vivid Dancers may be vulnerable to disturbance of their tiny scattered habitats. Spring-fed streamlets in dry forests and grasslands are often polluted and trampled by cattle. Most populations live in hot springs, which are almost always developed by people. Flight period: B.C., early May to mid October.

Emma's Dancer *Argia emma*

Male is violet; female is normally brown. The top of the thorax has a narrow black stripe less than half as wide as adjacent pale stripes. Length: ♂ 36 mm, ♀ 37 mm.

Range: Montane. In southern B.C., from the Fraser Valley to the Shuswap and Kettle River regions.

Field notes: Common at scattered localities along rivers, creeks and, sometimes, wave-washed lake beaches. The larvae lurk in debris and among plant stems in creek pools and under rocks in riffles. Flight period: B.C., early June to late September.

Red Damsels *Amphiagrion*

Restricted to North America; the only damselfly genus in B.C. whose adults are red in both sexes. Adults are robust, with stubby abdomens and short legs; they have a prominent tubercle on the underside of the thorax. In larvae, the hind corners of the head are pointed and turned outward. There are only two species, one eastern and one western.

Western Red Damsel *Amphiagrion abbreviatum*

Males are red, the thorax black on top. Females are all red or orange. Length: ♂ 26 mm, ♀ 27 mm.

Range: Western. In B.C., from the Prince George area south.

Field notes: Common in scattered localities at low and mid elevations, especially in southern valleys. Lives in marshy places with plenty of grasses and sedges: ponds, sloughs, spring-fed pools and slow streams. Adults fly close to the ground and perch frequently in low vegetation. Flight period: B.C., early May to early October.

Sprites *Nehalennia*

Small and delicate damselflies; five species live in the Americas and just one in Eurasia. The two Canadian species, unlike any others in the pond damsel family, are completely metallic green on top of the thorax; they also lack the pale spots behind the eyes that are common in many other members of the family.

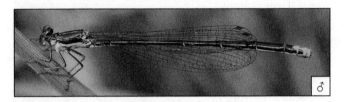

Sedge Sprite *Nehalennia irene*

The smallest and most delicate damselfly in B.C. The abdomen is dark with a blue tip; the top of the thorax is metallic green.
Length: ♂ 26 mm, ♀ 27 mm.

Range: Northern. Widespread in B.C.'s interior; on the coast, recorded rarely on Vancouver Island.

Field notes: Inconspicuous but common in suitable habitat; flies weakly in dense grasses and sedges. Most abundant in sedge meadows and lakes bordered by sedges. While laying eggs in floating plants, the female perches horizontally and the male, clasping her thorax with the tip of his abdomen, holds himself stiffly at a 45° angle. Flight period: B.C., early May to mid September (mostly in late spring and early summer).

Eurasian Bluets *Coenagrion*

There are two genera of bluets: *Coenagrion* and *Enallagma*. *Coenagrion* live mainly in Europe and Asia. Two species range across most of northern North America: the common Taiga Bluet and the rarer Subarctic Bluet. A third, the Prairie Bluet, flies on the Great Plains. Most Eurasian Bluet adults fly in late spring or early summer. They are similar to those of *Enallagma* — males are blue and black (but often green below); but the structure of the male appendages is different and females have no vulvar spine.

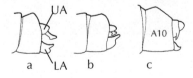

Figure 5. Male appendages of Eurasian Bluets, side view: a, Taiga; b, Subarctic; c, Prairie.
A10 = abdominal segment 10;
UA, LA = upper, lower appendages.

Taiga Bluet *Coenagrion resolutum*

Male's thorax has blue stripes on top, usually undivided; the underside is unmarked; abdominal segment 2 has a U-shaped black mark on top; appendages, fig. 5a. Female is blue to yellow-green or tan, marked with black; tops of segments 8 to 10 are black. Length: ♂ 29 mm, ♀ 30 mm.

Range: Northern. Widespread in most of B.C. and north to the Arctic treeline.

Field notes: Common in the Yukon and throughout B.C.'s interior, but mostly at higher elevations in the south; rare on the coast, it has not been found in the Lower Mainland or on Vancouver Island. Lives in many types of habitats, but is most common in sedge marshes and peatlands. Flight period: B.C., late May to early September; Yukon, late May to early August.

Subarctic Bluet *Coenagrion interrogatum*

The thorax has strong dark marks underneath; blue stripes on top are divided like a thick exclamation mark; abdominal segment 2 has a black mark on top and a large mark on each side; appendages, fig. 5b. Female is blue or greenish, marked with black; abdominal segment 8 is black at the base, segments 9 and 10 are mostly blue, and 9 has a black spot on top.
Length: ♂ 30 mm, ♀ 31 mm.

Range: Northern. In our region, widespread east of the Coast Mountains from the Arctic treeline to the U.S. border.

Field notes: The most boreal of North American damselflies, this species lives in a variety of wetlands, although it is common only in those containing floating aquatic moss. In southern B.C. it occurs only in subalpine wetlands in the mountains and high plateaus. Flight period: B.C., late May to late August; Yukon, late May to early August (uncommon after mid July).

Prairie Bluet   *Coenagrion angulatum*

The top of the thorax has undivided blue stripes. Abdominal segment 2 has a black mark on top, but usually none on the sides. Appendages, fig. 5c. Female yellow-green to tan, marked with black; the bases and sides of abdominal segment 8 are mostly pale, and the tops of segments 9 and 10 mostly black.
Length: ♂ 28 mm, ♀ 31 mm.

Range: Western. In B.C., east of the Rocky Mountains in the Peace River and Fort Nelson regions.

Field notes: Common at grassland ponds and forest marshes on the northern Great Plains and adjacent boreal forests. Flight period: B.C., late May to early August.

American Bluets   *Enallagma*

There are two genera of bluets: *Coenagrion* and *Enallagma*. *Enallagma* is a large genus, predominantly North American, containing most of the common blue-and-black damselflies in our region. Species identification requires close inspection. Females can be especially tricky: they are blue, green or brown; on most, the base of the abdominal segments is pale and the tips black; and all have a vulvar spine (fig. 7). Males are mostly sky blue with black stripes on the thorax and black rings around the abdominal segments; their appendages offer the best identification clues. The larvae, similar to those of *Coenagrion*, are patterned in brown or green and usually climb in aquatic vegetation.

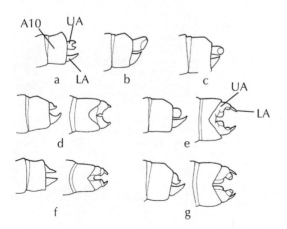

Figure 6. Male appendages for American Bluets: a-c, side view; d-g left, side view; d-g right, top view. a, Marsh; b, Familiar; c, Tule; d, Alkali; e, Boreal; f, Hagen's; g, Northern. A10 = abdominal segment 10; UA, LA = upper, lower appendages.

Tule Bluet   *Enallagma carunculatum*
Male has more black than blue on the middle abdominal segments (other species have more blue); appendages, fig. 6c. Female has a black stripe along the top of abdominal segment 8. Length: ♂ 32 mm, ♀ 33 mm.

Range: Southern. Widespread over at least the southern half of the province.

Field notes: Common at low and moderate elevations. In B.C., typical of lakes and large marshes rather than small ponds. Tolerates alkaline conditions and is often common along exposed lakeshores, especially where there are stands of bulrushes or cattails. Flight period: B.C. mid May to late October.

Familiar Bluet   *Enallagma civile*
Male's appendages, fig. 6b. Female has no pale area on top of segment 8. Length: ♂ 35 mm, ♀ 33 mm.

Range: Southern. In B.C., recorded only in the Cariboo region.

Field notes: One of the most common and widespread American damselflies, especially east of the Rockies, but recorded only once in B.C., an old record from Bridge Lake in the Cariboo. Closely related to the Tule Bluet. Often colonizes temporary and newly created water bodies. Flight period: B.C., (probably) early June to mid October.

Alkali Bluet   *Enallagma clausum*
Male's appendages, fig. 6d. All or most of female's segment 8 is pale.
Length: ♂ 35 mm, ♀ 34 mm.

Range: Western. Scattered localities in B.C.'s southern interior north to the Cariboo region.

Field notes: Prefers saline ponds and lakes in grasslands and dry forests. Flight period: B.C., mid June to early September.

Boreal Bluet *Enallagma boreale*

Male's appendages, fig. 6e (the view from above is most distinctive). The rear half of female's segment 8 is black, and a pale area at the base is sometimes divided by a black line. Length: ♂ 33 mm, ♀ 35 mm (largest in the far north).

Range: Northern. Throughout B.C. and north to the treeline in the Yukon.

Field notes: Lives in a wide variety of habitats – it swarms around warm saline ponds in dry grasslands, can be common along the marshy shores of lakes and ponds, and is a typical resident of cold peatland waters. Flight period: B.C., late April to mid October; Yukon, late May to mid September.

Northern Bluet *Enallagma cyathigerum*

Male's appendages, fig. 6g. The rear half of female's segment 8 is usually black on top, the pointed end dividing a pale area at the base. Length: ♂ 33 mm, ♀ 32 mm.

Range: Northern; also in northern Eurasia. Widespread throughout B.C. and the Yukon north to the Arctic treeline.

Field notes: The most common damselfly in southern B.C.; but in the Yukon, not as common as the Boreal Bluet. Lives mostly along the marshy shores of ponds and lakes, but usually avoiding acidic and saline waters. In B.C., adults usually emerge a week or two after those of its close relative, the Boreal Bluet. Flight period: B.C., early May to late October; Yukon, late May to early September.

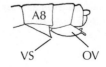

Figure 7. Northern Bluet female's abdomen tip (side view). A8 = abdominal segment 8; OV = ovipositor; VS = vulvar spine.

Hagen's Bluet *Enallagma hageni*
Male's appendages, fig. 6f. Female's segments 8 to 10 are entirely black on top. Length: ♂ 30 mm, ♀ 31 mm.

Range: Transition. In B.C., from the Kamloops area north to the Peace and Liard river drainages. Most often seen in the Cariboo and Prince George regions.

Field notes: An uncommon inhabitant of marshy lakes and ponds. Flight period: B.C., late May to early September.

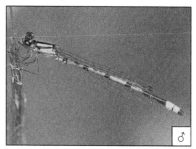

Marsh Bluet *Enallagma ebrium*
Male's upper appendages are forked, fig. 6a. Female's segments 8 to 10 are entirely black above. Length: ♂ 29 mm, ♀ 30 mm.

Range: Transition. In B.C., widespread east of the Coast Mountains, ranging north into the Peace and Liard regions.

Field notes: Common in the southern half of the province. Can be abundant, especially in marshes or open lakeshores. Usually avoids peatlands and other acidic habitats. Flight period: B.C., early June to mid September.

Forktails *Ischnura*
Found almost everywhere dragonflies live, though mostly absent from boreal habitats; distribution in North America is decidedly southern. B.C. has four species, but only the two most widespread – the Pacific Forktail and the Western Forktail – are encountered often. None are found in the Yukon. Male forktails in B.C. are mostly black, blue and green. The abdomen is black above and has a blue tip; the last segment bears a distinct forked projection on top, which gives the group its English name. Females may be the same colour as males or may have a tan, pink or orange thorax when immature; they may darken with extensive pruinescence as they age. Larvae are similar to those of bluets, but the gills usually have long, tapered tips.

Pond Damsels

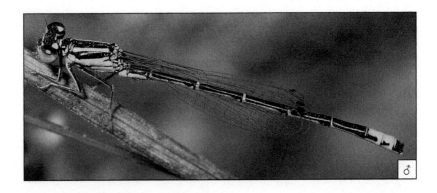

Western Forktail *Ischnura perparva*

Our smallest forktail. Male's thorax has green sides, and a dark top with green stripes; appendages, fig. 8c. Immature female's thorax has orange sides and stripes on top; the abdomen is black with an orange base. Mature female's pale areas darken to olive, but eventually, the whole body is covered by grey-white pruinescence. Females have no vulvar spine. Length: ♂ 26 mm, ♀ 27 mm.

Range: Western. Southern third of B.C. to about 52°N.

Field notes: Common at low and moderate elevations, but generally less common than the Pacific Forktail. Inhabits ponds and marshy lakeshores, and is more common in slowly flowing streams than the Pacific Forktail. The stocky little females lay eggs alone, their pruinose bodies and green eyes making them easy to identify. Flight period: B.C., early May to early October.

A mature pruinose Western Forktail.

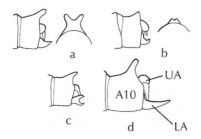

Figure 8. Forktail male appendages, side view; in *a* and *b*, the hind view of the top of abdominal segment 10 is on the right: a, Pacific; b, Plains; c, Western; d, Swift.
A10 = abdominal segment 10;
UA, LA = upper, lower appendages.

Pacific Forktail *Ischnura cervula*

Male's thorax has blue sides and a black top with two pairs of pale dots; appendages, fig. 8a. Female's thorax is sometimes coloured like the male's (dots elongated), but more often green or tan to pink with dark lines of variable thickness above; the abdomen has a blue tip. Most females darken with age; they often have a vulvar spine.
Length: ♂ 29 mm, ♀ 30 mm.

Range: Montane. Widespread in the southern half of B.C., to about 55°N.

Field notes: One of the most common damselflies of low and moderate elevations across southern B.C. Most common around marshes and the marshy edges of small lakes and ponds; the only dragonfly that often flies among dense cattails and bulrushes. Prefers neutral and alkaline waters. It has one of the longest flight seasons of any dragonfly in B.C., and is usually the first to appear in spring. Flight period: B.C., early April to late October.

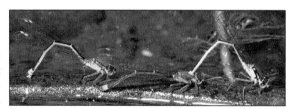

Female Pacific Forktails of different colour forms laying eggs.

Plains Forktail *Ischnura damula*

Male's coloration is like the Pacific Forktail's; appendages, fig. 8b. Female's thorax is often similar to the male's, or has pale spots on top joined into stripes; it may also be tan to pink. The abdomen's tip is blue. Females are less prone than Pacific Forktails to darken with pruinescence; they have a vulvar spine. Length: ♂ 28 mm, ♀ 27 mm.

Range: Western. In B.C., found only at Liard River Hot Springs on the Alaska Highway.

Field notes: The isolated population at Liard River Hot Springs is probably a relic of a more widespread distribution during a warmer climatic period. Closely related to the Pacific Forktail and similar in appearance, but in B.C., the ranges of the two species do not overlap. Flight period: B.C., late June to mid August.

Swift Forktail *Ischnura erratica*

Our largest forktail. Blue on the tip of the abdomen extends to segment 7. The thorax has blue sides and blue or green stripes on top; female's thorax is sometimes green or orange instead of blue. Male's appendages, fig. 8d. Length: ♂ ♀ 32 mm.

Range: Pacific Coastal. In B.C. only in the lowlands of the south coast.

Field notes: Lives around ponds, slow streams and lakes, including those in peatlands. Abundant in some locations, but not as common as Pacific or Western forktails. Flight period: B.C., late April to late August (mostly in May and June).

DRAGONFLIES   Suborder Anisoptera

Petaltails   Family Petaluridae
Abundant fossils show that the petaltails flourished in the Jurassic Period, at least 150 million years ago, well before the landmass that is now B.C. and the Yukon existed. Today, 11 species persist in widely scattered regions — New Zealand, Australia, Chile, Japan, the Appalachian Mountains and western North America. The eyes of petaltails are separated and the pterostigma is long and narrow; females have an ovipositor. Larvae live where water seeps and trickles over the ground, and many species make burrows; they are amphibious, spending much time out of water.

Black Petaltail   *Tanypteryx hageni*
The body is black with yellow spots, the face yellow with separated brown eyes. Male's upper appendages are flat and strongly angled outward. Female has a short, curved ovipositor. Length: ♂ 56 mm, ♀ 53 mm.

Range: Montane. In Canada, found only on B.C.'s mainland coast.

Field notes: Lives at mid to high elevations in the Cascade and southern Coast mountains, and at sea level on the central coast to about 53°N. Larvae burrow in mud and moss saturated by trickling water seeping from hillsides; many burrows can be concentrated in a small area. Adults perch on tree trunks, logs, rocks and the ground. They can be tame and frequently land on people. Flight period: B.C., early July to early September.

Darners   Family Aeshnidae
Large, swift-flying dragonflies, usually marked with blue, green or yellow. Adults hunt tirelessly for insects over ponds, lakes and streams, and wander widely in search of prey. Most species rest in a vertical position, but a few sit flat on the ground. Females have a prominent ovipositor and lay eggs in water plants or floating wood above or below the water line. Larvae are slender and sleek, with flat labia lacking bristles; they are rapacious hunters among water plants.

Darners

Mosaic Darners *Aeshna*

Mosaic Darners are common in B.C. and the Yukon; they fly everywhere dragonflies are found. All 13 species are large and can usually be distinguished by their variations on a basic colour pattern. Generally, the body is brown, and each side of the thorax has a pair of blue, green or yellow stripes – their shape is important in identification (see fig. 9). Look also for the colour of the face and the line across its middle. Viewed from above, the forehead bears a distinctive T-shaped mark, called the "T-spot" (fig. 10). The abdominal spots on males are usually blue, and on females green, yellow or blue. Male upper appendages come in three types (fig. 11).

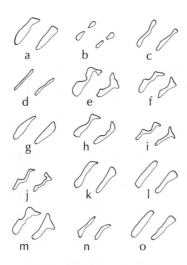

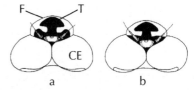

Figure 10. Mosaic Darners' T-spot, top view: a, Zigzag; b, Azure. CE = compund eye; F = face; T = T-spot.

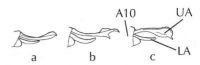

Figure 9. The shapes of the side stripes on the thorax of Mosaic Darners (the front stripe is on the left):
a, Black-tipped;
b, Variable – spotted form;
c, Variable – southeastern interior;
d, Variable – interior form;
e, Lake; f, Canada; g, Sedge;
h, Subarctic; i, Azure; j, Zigzag;
k, Shadow; l, Paddle-tailed;
m, Lance-tipped; n, California;
o, Blue-eyed.

Figure 11. The three types of appendages of male Mosaic Darners, side view:
a, **simple** (most species; Canada is shown) broadens slightly toward a rounded or pointed tip;
b, **forked** at the tip (Blue-eyed);
c, **flattened** (Lance-tipped, Paddle-tailed and Shadow) broadens and flattens toward the tip.
A10 = abdominal segment 10;
UA, LA = upper, lower appendages.

Lake Darner   *Aeshna eremita*

Large; looks much like the Canada Darner, except its face is yellow-green with a black line and it lacks spots under the abdomen. Thorax stripes are blue to green; shape, fig. 9e. Pale areas on females are usually green or yellow-green, but sometimes blue. Male's upper appendages simple (fig. 11a). Length: ♂ 75 mm, ♀ 73 mm.

Range: Northern. Widespread in B.C. and in the forested valleys of the southern Yukon; in the northern Yukon, it lives only in the low-lying Porcupine River system.

Field notes: The largest Mosaic Darner in Canada and one of the most often encountered dragonflies in the northern forests of North America. In southern B.C., it lives at all elevations, but is most common around forest lakes at mid and high elevations. Prefers lakeshores with little emergent vegetation; it also occurs in deep fens and bogs, and around lakes and ponds surrounded by sedges. It may fly early in the morning and in the evening when the temperature is cool and the light is low. It likes perching on tree trunks. Flight period: B.C., mid June to late October; Yukon, late June to early September.

Variable Darner
Aeshna interrupta

The scientific name comes from the distinctly broken thorax stripes, usually on males found on B.C.'s coast (fig. 9b); interior populations normally have complete narrow stripes (fig. 9c-d). This variation gives the species its English name. The stripes are usually yellow-green below and blue above. The face is green-yellow to pale blue with a black line. Pale areas on females are usually green or yellow, but some are blue. Male's appendages simple (fig. 11a).
Length: ♂ 70 mm, ♀ 66 mm.

Darners

Range: Northern. Widespread throughout B.C. and to the Porcupine River drainage in the Yukon.

Field notes: Common in B.C. and southern Yukon valleys; rare in the Porcupine basin. Lives in many habitats from northern and mountain peatlands to cattail marshes and temporary pools. It is the characteristic darner of grassland ponds. Flight period: B.C., late May (though uncommon before July) to early November; Yukon, late June to early September.

Canada Darner   *Aeshna canadensis*
Thorax stripes as in fig. 9f. The face is pale green; if a line is present, it is thin and pale brown. Blue spots mark the underside of the abdomen.

Male's thorax stripes are blue to green; upper appendages simple (fig. 11a). Female has pale green (sometimes blue) markings.
Length: ♂ 68 mm, ♀ 70 mm.

Range: Transition. Widespread B.C. and the southern Yukon.

Field notes: Not usually the predominant species at a locality, except on the south coast, where it can be abundant in some places. Most common in the south, rare in northern B.C. and extremely rare in the Yukon. Lives in peaty lakes, flooded beaver ponds, and sedge and cattail marshes at low and medium elevations. Flight period: B.C., mid June to late October; Yukon, late June to early September.

Black-tipped Darner   *Aeshna tuberculifera*
Large and slender. Thorax stripes are green to blue; shape, fig. 9a. Its face is green, the face line thin and pale brown or absent. Segment 10 of the abdomen is all black, as the English name indicates. Female's abdomen has blue spots, like the male's. Male's upper appendages simple (fig. 11a) with a tubercle underneath near the base; female's appendages are large.
Length: ♂ 73 mm, ♀ 75 mm.

Mosaic Darners

Range: Transition. Sparsely distributed across the moister regions of southern and central B.C.

Field notes: An uncommon dragonfly of peatland pools and peat-margined lakes. Unlike most mosaic darners, females patrol like males and often lay eggs in vegetation above the waterline. Presumably, this behaviour reduces the amount of attention that males give them, allowing more time for uninterrupted egg-laying. Flight period: B.C., mid June to early October.

Sedge Darner   *Aeshna juncea*

Thorax stripes are yellow-green below, blue above and bordered with black; shape, fig. 9g. The face is yellow to yellow-green with a black line. Pale spots mark the undersides of the abdominal segments. The pale areas on females are usually green or yellow-green, but on some they are blue. Male's upper appendages simple (fig. 11a).
Length: ♂ 66 mm, ♀ 65 mm.

Range: Northern; also across northern Eurasia. Widespread in B.C. and north almost to the treeline in the Yukon.

Field notes: Uncommon in southern valleys in B.C. More common at higher elevations and northward, it is the most common darner in the Yukon. Lives in a variety of habitats, mainly containing acidic waters, but it is most abundant in peatlands dominated by extensive stands of sedges. Flight period: B.C., mid June to early October; Yukon, late June to mid September.

Subarctic Darner
Aeshna subarctica

Similar to the Sedge Darner, but can be distinguished by the thorax stripes, which are shaped differently (fig. 9h) and lack a black border. Male's upper appendages simple (fig. 11a). Length: ♂ 68 mm, ♀ 66 mm.

Range: Northern; also across northern Eurasia. Widespread throughout B.C. and the Yukon.

Field notes: Lives in peatland habitats – bogs and deep fens that are dominated by aquatic mosses. Flight period: B.C., mid June to early October; Yukon, late June to mid September.

Zigzag Darner   *Aeshna sitchensis*

Small, similar to the Azure Darner, but its thorax stripes are yellow to blue, and the hind one is more zigzagged and less T-shaped (fig. 9j); the facial T-spot has a crescent-shaped base (fig. 10a). Male's abdomen has

large blue spots, though smaller than the Azure Darner's. The face is yellow to green. Female's abdominal spots are blue or yellow-green. Male's upper appendages simple (fig. 11a). Length: ♂ 59 mm, ♀ 56 mm.

Range: Northern. Widespread throughout our region.

Field notes: Restricted to specific peatland conditions, where it can be abundant: bogs or fens where the surface is mossy and sparsely vegetated with short, evenly spaced sedges, and where open water, if present at all, is reduced to small, shallow, mud- or moss-bottomed ponds and puddles. Adults perch on the ground, rocks and logs, or vertically, low on tree trunks. Flight period: B.C., late June to early October; Yukon, late June to late August.

Azure Darner   *Aeshna septentrionalis*

Small, similar to the Zigzag Darner. Thorax stripes are pale blue or green; shape, fig. 9i. The face is blue or green; T-spot, fig. 10b. Male's abdomen has more blue than the Zigzag Darner's, the blue spots usually fused to form irregular stripes; upper appendages simple (fig. 11a). Female has blue or yellow-green abdominal spots. Length: ♂ 60 mm, ♀ 57 mm.

Range: Northern. Widespread in the Yukon. In B.C., it ranges south only in the mountains and on northern plateaus to about 51°N in the Coast and Rocky mountains.

Field notes: The most boreal of our darners. In the Yukon, it is the only dragonfly known to breed north of the British Mountains; uncommon in the southern valleys. Yukon records are from a variety of peatlands, marshes and ponds, many with floating aquatic mosses. In B.C., it lives in subalpine peatlands with ponds. Perches low on tree trunks, or on stones, logs or moss. Flight period: B.C., mid July to mid September; Yukon, mid June to mid September.

Paddle-tailed Darner *Aeshna palmata*

The face is greenish yellow with a black line; the rear of the head is black. Thorax stripes are almost straight (fig. 91). The abdomen has pale spots on top of segment 10, but none underneath on any segments. Male's stripes are yellow below and green to blue above, and its abdominal spots are blue. Females can be coloured like the male, but most have green-yellow thorax stripes and abdominal spots. Male's upper appendages flattened (fig. 11c). Length: ♂ 72 mm, ♀ 67 mm.

Range: Montane. Widespread throughout B.C.; restricted to southern valleys in the Yukon.

Field notes: One of the most frequently encountered and abundant dragonflies in B.C., though not as plentiful in the far north. Lives in a wide range of habitats, but typically in lakes and ponds in or near woodlands. Flight period: B.C., mid May (usually late June) to early November.

Shadow Darner *Aeshna umbrosa*

Similar to the Paddle-tailed Darner, but darker and more slender. Thorax stripes are yellow to green, often blue above and outlined with dark brown; shape, fig. 9k. The face is pale green with a pale brown line or no line; the rear of the head is partly pale. Spots on top of the abdominal segments are green or blue; segment 10 is usually black on top. The underside of the abdomen has pale blue spots. The pale areas on females are green, yellow or, rarely, blue. Male's upper appendages flattened (fig. 11c).
Length: ♂ 70 mm, ♀ 68 mm.

Range: Transition. Widespread in B.C. and the southern Yukon.

Field notes: More common in the south; rare in the southern Yukon. Partial to forest lakes and slow-moving streams; as its name suggests, it likes shady habitats. Often found alongside the Paddle-tailed Darner, its close and usually more common relative. The Shadow Darner is one of the latest flying species in B.C., especially in the south. Flight period: B.C., mid June to mid November.

Lance-tipped Darner *Aeshna constricta*

Thorax stripes usually yellow-green below and blue above (but on some females are all yellow); shape, fig. 9m. The face is pale green with a thin, pale brown line, or no line at all. Females have blue, green or yellow abdominal spots; segment 9 is longer than segment 8; the ovipositor and appendages are unusually large, the former extending well past the end of segment 9. Male's upper appendages flattened (fig. 11c).
Length: ♂ 71 mm, ♀ 69 mm.

Range: Transition. In the valleys of the southern interior, north to about 51°N.

Field notes: Rare at small ponds and open, warm, nutrient-rich marshes dominated by cattails and bulrushes; sometimes develops in waters that dry up in summer. Its preference for habitats that are often threatened by

Lance-tipped Darners   California Darners

human development make it vulnerable. Female Lance-tipped Darners, like their Black-tipped counterparts, lay eggs in vegetation well above the waterline. Flight period: B.C., early July to mid October.

California Darner   *Aeshna californica*

Thorax stripes straight (fig. 9n) and bordered with black. The eyes are blue and the face is pale blue with a black line. Male's thorax stripes are pale blue; its upper appendages simple (fig. 11a). Female's thorax stripes and abdomen spots are pale blue or yellow. Length: ♂ 60 mm, ♀ 59 mm.

Range: Montane. In B.C., south of about 52°N.

Field notes: Common around ponds, lakes and marshes at low and medium elevations. For a darner, remarkable for its springtime flight season. At the most southerly locations, it may appear in April, with the earliest dragonflies; by early August it is uncommon, just when many darners are reaching their peak abundance. Flight period: B.C., mid April to mid August.

Darners

Blue-eyed Darner
Aeshna multicolor

Thorax stripes straight (fig. 9o). Its face line, if it has one, is thin and pale brown. Male's face and eyes are sky-blue; the thorax stripes and abdomen spots are blue; upper appendages forked (fig. 11b). Female's thorax stripes and abdomen spots are either blue or yellow. Length: ♂ 69 mm, ♀ 67 mm.

Range: Western. In B.C., as far north as McBride, but widespread only south of about 51°N.

Field notes: Common around marshes and marshy lakeshores and ponds in southern valleys. Flight period: B.C., mid May to mid October.

Green Darners   *Anax*

Four species of this cosmopolitan genus live in North America, but only one reaches B.C. from the south.

Common Green Darner   *Anax junius*

One of our largest dragonflies, with a wingspan of almost 12 cm. The thorax is unmarked and green. Viewed from above, the forehead bears a black spot in a blue patch. The abdomen has a dark centre stripe; on immatures, the sides are reddish, but with age the male's become blue and the female's grey-green to violet. Length: ♂ ♀ 75 mm.

immature ♀

Range: Southern; also in parts of Asia and on Pacific islands. Southern B.C. to the Prince George region.

Field notes: Uncommon across southern B.C., but common in some coastal localities. Develops in warm marshes and ponds at low elevations. Southern Canada appears to have two populations — one seems to migrate north in the spring and lays eggs, their offspring developing rapidly and flying south in August and September; the other population is resident all year round, the adults taking flight in June and July. Flight period: B.C., late April to early October.

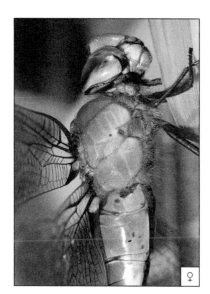

Clubtails Family Gomphidae

A large family, but poorly represented in our region. B.C. has only six species and the Yukon appears to have none (although two species may live in the southeast). Compared to some other families, clubtails are not common here, but they are easily recognizable by their widely separated eyes and their green or yellow bodies striped in brown and black. The tip of the abdomen, especially in males, is enlarged, giving them their English name. Females lack an ovipositor and drop their eggs directly into clear streams and along the sandy shallows of larger lakes; they lay their eggs without the protection of their mates. Larvae burrow in the bottom sediments of these water bodies.

Snaketails *Ophiogomphus*

Most snaketails live in eastern North America. The three species in B.C. fly along clear streams and lakeshores; uncommon in settled areas, at least partly because the burrowing larvae are sensitive to changes in water flow and siltation, and they are especially affected by poor logging practices. The mature adult's thorax is green; younger adults are more yellow. The abdomen has white or yellow marks on the sides and yellow spots on top; this snake-like pattern gives the genus its English and scientific names (*ophio* is Greek for "snake" and *gomphos* means "bolt" or "arrow"). The hindwings have a three-celled anal loop (fig. 2a). Male's upper appendages are short and usually pointed in side view.

Pale Snaketail *Ophiogomphus severus*

The face is yellow with no stripes. The thorax is green-yellow to green with fine brown stripes on the sides and a prominent oval spot in front of the forewing base. The abdomen is dark brown, sometimes mostly yellow below, with large yellow spots on top. Length: ♂ 51 mm, ♀ 52 mm.

Range: Western. Widespread in B.C. east of the Coast Mountains; it may occur in the southeastern Yukon, but has not yet been found there.

Field notes: Uncommon. Most records are from clear streams, but this species also develops in large, muddy rivers and in lakes, where it sometimes emerges on floating waterlily leaves. Flight period: B.C., early June to mid September.

♂ Pale Snaketail

♂ Boreal Snaketail

Boreal Snaketail *Ophiogomphus colubrinus*

The face is green with black stripes. The thorax is bright green with brown stripes, including one on the side behind the forewing base. Length: ♂ ♀ 47 mm.

Range: Northern. Widespread in northern B.C., and south to the Cariboo; it may occur in the southeastern Yukon, but it has not been found there yet.

Field notes: The most northern of all North American clubtails and the only transcontinental clubtail inhabiting boreal forests. Uncommon, but can be abundant in the right habitat – clear, warm streams flowing out of forest lakes. Larvae burrow in the sand and gravel on the stream bottom; adults perch on the ground or on streamside vegetation. Flight period: B.C., mid June to early September.

Sinuous Snaketail *Ophiogomphus occidentis*

Coloured much like the Boreal Snaketail, except its face is yellow-green with no stripes and its thorax has no stripe behind the forewing base (though there is sometimes a thin line on the side in front of the hindwing base). A conspicuous brown stripe in front of the forewing base is divided by wavy pale line – this gives the species its English name. Length: ♂ 50 mm, ♀ 51 mm.

Range: Montane. Widespread in B.C. south of about 51°N.

Field notes: Uncommon denizen of sunny stream banks and sandy lakeshore beaches at low elevations. The only clubtail on Vancouver Island, where the best place to see it is along the Nanaimo River. Flight period: B.C., early June to early October.

Grappletail *Octogomphus*

This genus contains a single distinctive species restricted to the Pacific coast forests of North America from British Columbia to Baja California. The genus name comes from the eight points of the male's appendages: the uppers are broad and forked, the lower divided into four points. The hooked outer branches of these claspers give the species its English name.

Grappletail *Octogomphus specularis*

The face is yellow-green striped with black. The top and sides of the thorax are pale green to green-yellow with a broad black stripe between them. The abdomen is nearly all black, and the male's is widest at segment 10. Length: ♂ 49 mm, ♀ 50 mm.

Range: Pacific Coastal. In B.C., found only in the lower Fraser Valley, where it reaches the northern limit of its range.

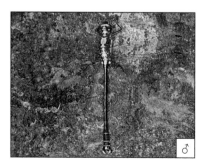

Field notes: Lives along wooded streams draining lakes. Adults perch on rocks along the stream or on trees and bushes away from water. Flight period: B.C., early June to late September.

Common Clubtails *Gomphus*
A large genus, especially numerous in eastern North America, but only a single, rare species enters B.C. in the southern interior.

Pronghorn Clubtail *Gomphus graslinellus*
The face is green; the thorax is grey-green to yellow, with two brown stripes of about equal width (both stripes are often divided into pairs of thinner ones); the abdomen is black with a yellow stripe on top of each segment, and yellow spots on the sides, larger near the tip. The English name comes from the shape of the male's upper appendages: like the horn of a Pronghorn antelope. Length: ♂ 48 mm, ♀ 50 mm.

Range: Transition. In B.C., found only in southern interior valleys.

Field notes: A rare species of warm lakeshores and streams. Larvae burrow in sand along wave-washed shores, especially where mats of water weeds grow offshore; when ready to emerge, they crawl only a few centimetres up the beach to metamorphose. Adults bask on sunny ground near water. When disturbed, they have a characteristic undulating flight. Flight period: B.C., late May to early August.

Hanging Clubtails *Stylurus*
Closely related to Common Clubtails, but fly later in summer and fall. Only one of twelve North American species lives in B.C., a rare inhabitant of the southern interior. Males often patrol back and forth over the water, then land on leaves and twigs that bend under their weight, leaving them hanging nearly vertically.

Olive Clubtail *Stylurus olivaceus*
Named for its colour: green face with a dark line, and grey-green thorax. No side stripes on the thorax, but broad, partly divided shoulder stripes and a prominent triangular patch on top. The abdomen is black with a yellow stripe or spot on top of each segment and yellow on

the sides, more extensive near the tip. The male's upper appendages are pointed. Length: ♂ 53 mm, ♀ 55 mm.

Range: Montane. In B.C., found only in the Thompson-Okanagan and Boundary regions.

Field notes: Rare in dry interior valleys. Breeds along warm streams and lakeshores with sandy or muddy edges. Adults fly over the water, landing along the shore or on shoreline vegetation. Flight period: B.C., mid July to mid October.

Spiketails   Family Cordulegastridae

Large black-and-yellow dragonflies with eyes meeting at a single point on top of the head. Spiketails live along streams where the males patrol steadily up and down for long distances. The English name comes from the female's long, spike-like ovipositor, used for placing eggs in streambeds. The large, squat, hairy larvae bury themselves in the sediment to await their prey.

Pacific Spiketail   *Cordulegaster dorsalis*

The face is yellow with a dark line; blue eyes barely meet at a point on top of the head. The body is black, with yellow stripes on the thorax and yellow spots on top of the abdomen. Length: ♂ 75 mm, ♀ 80 mm.

Range: Montane. Coastal valleys; southern interior to about 51°N.

Field notes: One of our largest dragonflies. Fairly common along small coastal woodland streams flowing from lakes; rare east of the Coast Mountains in southern B.C., and where it does occur, it flies along small warm streams, especially ones fed by springs. Egg-laying females hover vertically, moving up and down, shoving eggs into the sand and silt of the streambed. Perches vertically. Flight period: B.C., mid May to early September (most are seen in July).

Western River Cruiser

Cruisers   Family Macromiidae
Large brown-and-yellow dragonflies of rivers and wave-washed lakeshores, where the adults fly out over the water. The thorax is encircled between the wings by an oblique yellow band. Larvae sprawl on the bottom silt and sand; they have long spider-like legs and a horn-like projection between the eyes.

Western River Cruiser   *Macromia magnifica*
The face is yellow with dark brown marks. Its grey eyes meet broadly on top of the head. The body is dark brown with a single yellow stripe on the side of the thorax and yellow bands on top of the abdominal segments. The thorax is faintly metallic, thinly pruinose with age. The legs are unusually long. Populations around Shuswap Lake and in the Fraser Valley are darker than those from drier valleys – the yellow bands on abdominal segments 3 to 6 are reduced to small, divided spots; they were once considered a separate species, *Macromia rickeri*.
Length: ♂ 69 mm, ♀ 71 mm.

Range: Montane. Fraser Valley; southern interior valleys south of 51°N; not recorded in the Columbia or Kootenay river valleys.

Field notes: Uncommon along warm lake margins and sandy rivers. Males patrol with a swift, direct flight, low over the water, but like other large dragonflies, cruisers often feed in open areas far from water. Females lay eggs alone, striking their abdomens on the water surface. Both sexes perch vertically on vegetation. Flight period: B.C., early June to late September.

Emeralds   Family Corduliidae
Medium-sized dragonflies most often seen around lakes, boggy streams and peatlands in the mountains or in the north. Of 16 species in our region, 13 have Northern or Beringian ranges. The eyes, often brilliant green, meet broadly on top of the head. The shape of the anal loop in the hindwing is distinctive (fig. 2b). Adults seldom perch during feeding and males frequently hover when patrolling for mates; when resting, they normally hang vertically or obliquely from vegetation. In flight, a male frequently arches its abdomen, which is often narrower at the base and tip. Larvae are usually squat and rather hairy; they sprawl in the mud and detritus in the bottom of the waters where they live.

Baskettails   *Epitheca*
Baskettails live around the northern hemisphere. Rather than metallic green and black, like other emeralds, baskettails have a brown thorax and a dark abdomen with yellow marks on the sides. The hindwings are marked with brown at the base. Females fly with the end of the abdomen curled upwards, the forked, finger-like vulvar lamina holding a ball of eggs as in a basket, which gives the group its English name. To lay the eggs, a female dips the egg mass into the water and it uncoils in long, gelatinous strands that float near the surface. Many females may contribute to communal egg masses. Larvae are less hairy and less coated with algae than those of our other emeralds; they have prominent dorsal and lateral spines on a broad abdomen. Some biologists place the North American species in the genus *Tetragoneuria*.

Spiny Baskettail   *Epitheca spinigera*
The orange-yellow face has a dark T-spot. Male's appendages, fig 12a. Female's appendages are 3 mm long; vulvar lamina, fig. 12c.
Length: ♂ ♀ 48 mm.

Range: Transition. Widespread in southern B.C.

Field notes: Infrequently encountered in the southern valleys of the province, although at times it can be numerous. Breeds in lakes and ponds, but often seen hunting well away from water. Flies early in the season. Flight records: B.C., early May to late August (most in June).

Beaverpond Baskettail
Epitheca canis

The face is orange-yellow with no T-spot. Male's appendages, fig. 12b. Female's appendages are 2 mm long; vulvar lamina, fig 12d. Mature females have an extensive brown wash to their wings.
Length: ♂ ♀ 45 mm.

Range: Transition. Scattered records from the south coast to the Peace River drainage; no records from the dry southern valleys of the interior.

Field notes: Rare inhabitant of marshy lakeshores, boggy ponds and backwaters of slowly flowing streams. A spring and early summer species. Flight period: B.C., early May to mid August.

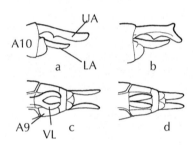

Figure 12. Identifying features for Baskettails.
Above, male appendages, side view: a, Spiny; b, Beaverpond.
Below, female vulvar lamina, view from below: c, Spiny; d, Beaverpond.
A9, A10 = abdominal segments 9, 10; UA, LA = upper, lower appendages; VL = vulvar lamina.

Striped Emeralds   *Somatochlora*

The scientific name comes from two Greek words: *soma*, meaning "body", and *khloros*, meaning "green". The English name refers to the yellow or white markings on the sides of the metallic green or bronzy thorax, but most of the distinctly striped species live in eastern North America; western species have spots, short bars or no marks at all. They all have brilliant green eyes. Most North American species are boreal and Appalachian; 13 of 26 occur in our region. Most live around northern or mountain lakes and peatlands.

Species can be hard to identify, especially the females. Look for pale marks on the side of the thorax and white rings between the segments of the abdomen. Hudsonian, Ringed and Lake emeralds have narrow, white abdominal rings. Examine the shapes of the male's upper appendages (fig. 13) and the female's vulvar lamina (fig. 14). And look for a brown spot at the base of the hindwings – not the membranule, which can also be dark, but an additional spot (Delicate, Muskeg and Whitehouse's emeralds have it; see fig. 15).

Brush-tipped Emerald
Somatochlora walshii

The side of the thorax has two pale marks, a short stripe at the front and an oval spot behind. The short, dark abdomen has small yellow spots on the sides of segments 5 to 7, at least. Male's appendages, fig. 13a. Female's vulvar lamina is dark; shape, fig. 14b.
Length: ♂ 44 mm, ♀ 47 mm.

Range: Northern. Throughout the southern two-thirds of B.C. to at least 56°N.

Field notes: Uncommon. Breeds in sedge meadows or fens, typically with small slow streams flowing through them. Flight period: B.C., late June to mid September.

Ocellated Emerald *Somatochlora minor*
The English name refers to two yellow eye-like spots on each side of the thorax. Otherwise, this species is scarcely marked, though the short, dark abdomen has some pale spots at the base. Male's appendages, fig. 13b. Female's vulvar lamina is long, fig. 14a. Length: ♂ 43 mm, ♀ 46 mm.

Range: Northern. Throughout B.C. to the valleys of the southern Yukon.

Field notes: Uncommon inhabitant of warm streams, both gently and rapidly flowing ones, in peatlands and forests. The stream bottom can contain any material from rocks to soft mud and organic matter. Females have a more conspicuous vulvar lamina than other emeralds; they use this spout-like egg-laying device to tap eggs into the water or into moss on stream banks. Flight records: B.C., early June to mid September; Yukon, mid June to late August.

Emeralds

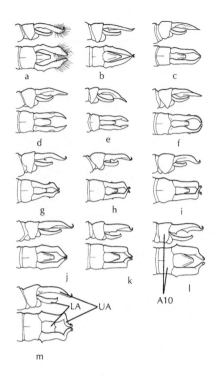

Figure 13. Striped emerald male appendages (side view above, top view below): a, Brush tipped; b, Ocellated; c, Delicate; d, Kennedy's; e, Forcipate; f, Mountain; g, Whitehouse's; h, Muskeg; i, Treeline; j, Quebec; k, Ringed; l, Hudsonian; m, Lake. A10 = abdominal segment 10; UA, LA = upper, lower appendages.

The upper appendages come in three main shapes: a-b, converging and the tips turned gently upward; c-f, pincer-like and the tips pointed downward with no curl at the tip; g-m, tips bent abruptly inward and strongly curled.

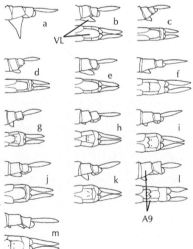

Figure 14. Female striped emeralds have a spout-like or flap-like vulvar lamina (side view above, top view below): a (side view only), Ocellated; b, Brush tipped; c, Delicate; d, Kennedy's; e, Forcipate; f, Mountain; g, Whitehouse's; h, Muskeg; i, Treeline; j, Quebec; k, Hudsonian; l, Ringed; m, Lake. A9 = abdominal segment 9; VL = vulvar lamina.

Figure 15. The hindwing of a Delicate Emerald. M = membranule; S = spot. Muskeg and Whitehouse's emeralds have a similar spot.

Delicate Emerald   *Somatochlora franklini*

Named for its slender body – obvious in males. The long, narrow abdomen is widest at segment 9; the side of the thorax has one pale spot, often obscured with age. The wings are short, just 3/4 the length of abdomen, and the hindwings have a dark spot at the base (fig. 15). Male's appendages, fig. 13c. Female's vulvar lamina, fig. 14c. Length: ♂ 50 mm, ♀ 45 mm.

Range: Northern. From about 51°N in central B.C. to the Porcupine River basin in the northern Yukon. Absent from the coast.

Field notes: A northerner, becoming scarcer in the southern parts of its range, where it lives at higher elevations. Usually flies with the Zigzag Darner, but rarer than the darner, at least in the south. Can be common in the right habitat. Prefers shallow, moss-bottomed bogs and fens evenly vegetated with sedges and horsetails; in these habitats, open patches of water are usually not visible from a distance. Males hover frequently, pivoting while stationary. Flight period: B.C., mid June to early September; Yukon, mid June to late August.

Kennedy's Emerald   *Somatochlora kennedyi*

Lacks conspicuous markings, though it may have one faint stripe on the sides of the thorax; usually no spots on the abdomen sides after segment 3. The abdomen is long, the male's widest at segment 6. Male's appendages, fig. 13d. Female's vulvar lamina is yellow; shape, fig. 14d. Length: ♂ 44 mm, ♀ 43 mm.

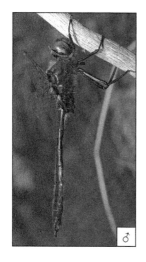

Range: Northern. From northeastern B.C. to the Porcupine River drainage in the northern Yukon.

Field notes: Uncommon in the Yukon, rare in northern B.C. Found in sedge fens and marshes where females lay eggs in shallow, open pools underlain by aquatic moss. Named for the American dragonfly researcher, Clarence Kennedy. Flight period: B.C. and Yukon, mid June to late August.

Forcipate Emerald   *Somatochlora forcipata*

Similar to the Mountain Emerald but more slender. The thorax sides have two whitish oval spots. The abdomen has small, dull-yellow spots (on segments 5 to 8 on males, and 3 to 7 on females), which are usually absent on Mountain Emeralds. Male's appendages, fig. 13e. Female's vulvar lamina is yellow; shape, fig. 14e. Length: ♂ 48 mm, ♀ 49 mm.

Range: Transition. From Kootenay National Park in the southern Rocky Mountains to Williston Lake; central B.C. plateaus.

Field notes: Rare; first discovered in B.C. in 1998. Lives around shallow spring-fed streamlets trickling through subalpine hillside fens, or in small pools associated with flowing groundwater in such situations. Females lay eggs in moss and algal mats. Patrolling males often fly in shady glades in open spruce forests. Flight period: B.C., mid June to early September.

Mountain Emerald   *Somatochlora semicircularis*

The sides of the thorax are metallic green with two oval yellow spots, the larger in front. The abdomen sides sometimes have small, dull yellow spots on segments 5 to 8. Male's appendages, fig. 13f (the shape gives this species its scientific name, *semicircularis*). Female's vulvar lamina, fig. 14f. Length: ♂ ♀ 50 mm.

Range: Montane. Widespread in B.C., and north to valleys in the extreme southern Yukon.

Field notes: The most common striped emerald in B.C., much more common in the south than in the north; rare in the southern Yukon. Thrives in sedge marshes and sedge-lined lakes and ponds, especially where the plants are tall and dense; males patrol and hover over these sedge beds. Also lives in fens and bogs. Flight period: B.C., late May to early October; Yukon, mid June to early September.

Whitehouse's Emerald   *Somatochlora whitehousei*

Similar to the Muskeg Emerald, but much more common. Both have a dark spot at the base of each hindwing (see fig. 15). Both have brassy green thorax sides with a short, indistinct yellow-brown stripe below the base of the forewing (sometimes an indistinct yellowish patch below the base of the hindwing). The difference is in the shapes of the male appendages (fig. 13g) and vulvar lamina (fig. 14g). Length: ♂ ♀ 47 mm.

Striped Emeralds

Above: the Mountain Emerald is the most common striped emerald in our region.

Right: Whitehouse's Emerald is strikingly similar to the rare Muskeg Emerald, which is not shown in this book.

Range: Northern. Widespread east of the Coast Mountains from southern B.C. to the Porcupine River drainage in the northern Yukon.

Field notes: Uncommon overall, but often abundant in suitable habitats; rare south of 51°N, where it lives in subalpine-forest peatlands. Prefers level, shallow, mossy fens and bogs dominated by sedges and buckbean. Males patrol across and around pools or small shallow puddles. Named for Francis Whitehouse, a B.C. banker and student of dragonflies. Flight period: B.C., early June to early September; Yukon, early June to late August.

Muskeg Emerald *Somatochlora septentrionalis*
Similar to Whitehouse's Emerald (see photograph on p. 65), but less common. Both have a dark spot at the base of each hindwing (see fig. 15). Both have brassy green thorax sides with a short, indistinct yellow-brown stripe below the base of the forewing (sometimes an indistinct yellowish patch below the base of the hindwing). The difference is in the shapes of the male's appendages (fig. 13h) and female's vulvar lamina (fig. 14h). Length: ♂ 43 mm, ♀ 44 mm.

Range: Northern. From about 54°N on the plateaus of central B.C. and about 52°N in the Coast Mountains to the southern Yukon.

Field notes: Rare. Lives in bogs and fens where males patrol shallow ponds and pools. Typically, these pools are set in sphagnum and other mosses and are characterized by soft, mucky margins and scattered sedges. Males fly irregularly over the open water and around the pool edges, hovering infrequently. Flight period: B.C. and Yukon, mid June to late August.

Treeline Emerald *Somatochlora sahlbergi*
Closely related to the Hudsonian and Ringed emeralds, but lacks the abdominal rings. Similar also to the Quebec Emerald, but their ranges do not overlap. The thorax sides are coppery-green, usually without obvious marks. Male's appendages, fig. 13i. Female's vulvar lamina, fig. 14i. (Hybrids with Ringed and Hudsonian emeralds can confuse identification; their sizes, abdominal rings and appendage shapes are intermediate between those of the parent species.) Length: ♂ 50 mm, ♀ 48 mm.

Range: East Beringian, and across northern Eurasia. Northern Yukon from the Blackstone River to the Porcupine River drainage.

Field notes: Has the most northerly breeding range of any dragonfly: as its name suggests, the Treeline Emerald lives within 100 km of the Arctic treeline and within 300 metres of the altitudinal treeline. It inhabits a range of ponds and pools, from roadside ditches to moss-margined lakes. All habitats have two main characteristics: aquatic mosses as the dominant vegetation, and deep, cold water. Males and females tend to fly out over the water rather than circle the pond margins as Hudsonian and Ringed emeralds do. The Treeline Emerald interbreeds with both these relatives where they occur together. Flight period: Yukon, late June to late August.

Ringed Emerald

♂ Treeline Emerald

Ringed Emerald   *Somatochlora albicincta*
Similar to the Hudsonian Emerald, but smaller. The thorax sides are brassy green and marked with a white bar, its ends pointed. The English and scientific names refer to narrow white rings between the segments of the abdomen (*albicincta* means "girdled with white"); the top of segment 10 has a pair of pale spots. Male's appendages, fig. 13k. Female's vulvar lamina, fig. 14l. Length: ♂ 50 mm, ♀ 50 mm.

Range: Northern. Widespread in B.C.; the only white-ringed emerald west of the Coast Mountains. In the Yukon as far north as the Porcupine River drainage.

Field notes: Patrols margins of firm-edged peatland ponds and slow streams, and the open, peaty margins of forest lakes. Its habitats usually contain relatively open, poorly vegetated, shallow water. Specimens on the B.C. coast are larger than usual; some on the Queen Charlotte Islands can be as large as the Lake Emerald. In the northern Yukon, in places where Ringed and Treeline emeralds fly together, these species often interbreed. Flight period: B.C., mid June to mid October; Yukon, mid June to mid September.

Quebec Emerald
Somatochlora brevicincta

Similar to the Ringed Emerald, but its abdomen lacks rings or they are reduced to small spots on the sides between segments 3 to 7; and the top of segment 10 is black. The thorax sides are brassy green with a white bar, its ends pointed. Female has reddish hairs on the hind margin of its head. Male's appendages, fig. 13j. Female's vulvar lamina, fig. 14j. Length: ♂ 48 mm, ♀ 45 mm.

Range: Northern. Central Rocky and Cariboo mountains near McBride to Williston Lake; central B.C. plateaus.

Field notes: Rare. First discovered in B.C. in 2000, several thousand kilometres west of Quebec (where it was originally discovered) and the rest of its range in the Atlantic provinces and Maine. In B.C., this species prefers level, mossy fens and bogs with shallow pools dominated by short sedges. It flies with Whitehouse's Emerald in these habitats, but is much less abundant. Flight period: B.C., mid June to early September.

Hudsonian Emerald   *Somatochlora hudsonica*

Very similar to the Ringed Emerald, but larger. The thorax sides are brassy green with one tan, blunt-tipped bar that can be indistinct. The abdomen has narrow white rings. Male's appendages, fig. 13l. Female's vulvar lamina, fig. 14k. Length: ♂ 52 mm, ♀ 52 mm.

Range: Northern; west of Hudson Bay only. Widespread in B.C. east of the Coast Mountains; north in the valleys of the Yukon to the Porcupine River drainage.

Field notes: Uncommon in most of its B.C. range, especially in the south, where it lives at higher elevations. Breeds in deep, sedge-bordered lakes and ponds and in peatland ponds. The males patrol the outer edge of the emergent vegetation. In the northern Yukon, in places where Hudsonian and Treeline emeralds fly together, these species often interbreed. Flight period: B.C., early June to mid September; Yukon, early June to mid August.

Lake Emerald

Lake Emerald *Somatochlora cingulata*
The largest emerald in our region, as big as a small darner. Similar to Hudsonian and Ringed emeralds, but larger. The thorax sides are dark brassy green-brown without marks; the abdomen has narrow white rings. Male's appendages, fig. 13m. Female's vulvar lamina, fig. 14m. Length: ♂ 60 mm, ♀ 58 mm.

Range: Northern. Southern and central B.C. east of the Coast Mountains.

Field notes: Often hard to find, but can be common around midday in its preferred habitat: mountain and plateau fishing lakes, usually edged with sedges and peat and dotted with floating waterlilies. Adults are often difficult to observe and catch, for they usually fly rapidly over open water several metres from shore. Flight period: B.C., late June to early September.

Hudsonian Emerald

Common Emeralds   *Cordulia*

Remarkably similar to their close relatives, Striped Emeralds (a much larger genus). There are only two species of Common Emeralds – ours and the Downy Emerald (*Cordulia aenea*) in Eurasia – but as the English genus name suggests, they are widespread and common.

American Emerald   *Cordulia shurtleffii*

The face is dark with brilliant green eyes. The thorax is metallic green-bronze and, unlike those of many Striped Emeralds, has no pale marks on the sides. The abdomen is black after segment 4 and the male's is widest at segment 8. Male's appendages are forked in views from below and the side, fig. 16a. Female's vulvar lamina, fig. 16b.
Length: ♂ 45mm, ♀ 46mm.

Range: Northern. Widespread in B.C. and the Yukon.

Field notes: The most commonly seen emerald in our region. Flies early in the season, usually well before Striped Emeralds appear, but both genera fly together later in the summer. Males patrol energetically and aggressively around forest lakes and peatlands, chasing off males of their own and other species. Flight period: B.C., early May to early September; Yukon, late May to early August.

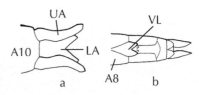

Figure 16. American Emerald: a, male appendages, view from above; b, female vulvar lamina, view from below.
A8, A10 = abdominal segments 8, 10; UA, LA = upper, lower appendages; VL = vulvar lamina.

Skimmers   Family Libellulidae
The largest dragonfly family in our region – 23 species in five genera live here and 2 species in two other genera are visitors. They come in many sizes and colours, many with bold wing markings or coloured veins. Their eyes meet broadly on top of the head. The anal loop in the hindwing is distinctive: foot-shaped with a long toe (fig. 2c). Most common around ponds, marshy lakeshores and sluggish streams, the adults dart about and most species spend a lot of time perched horizontally in the sun. Females lay eggs alone or in the company of guarding males. Most dip the tip of their abdomen into the water when releasing the eggs, but some will tap or splash the eggs into wet mud or moss, or simply flick them into a dry pond basin. Some larvae, like those of the emeralds, move sluggishly or squat on the bottom mud; others climb in vegetation.

King Skimmers   *Libellula*
Five striking king skimmers live in B.C., but only one ranges north to the Yukon. Most have banded or spotted wings, and in some species, males sport abdomens covered with white or bluish pruinescence. Showy and aggressive, king skimmers perch, hover and skim over the waters of ponds, lakeshores and sluggish streams. During egg laying, a female taps the water with the end of her abdomen; she flies alone or is guarded by her mate hovering nearby. Females of most species have small flaps on the sides of segment 8 that help them splash water along with their eggs, often depositing them on the shore.

Chalk-fronted Corporal

Chalk-fronted Corporal   *Libellula julia*
Mostly brown. Named "corporal" because immatures have a pair of pale stripes on the top of the thorax; they also have a dark stripe on top of the abdomen. Mature adults are "chalk-fronted" with distinctive pale pruinescence on the front of the thorax, as well as the base of abdomen –

white on males, grey on females. The wings are clear except for small, dark marks at the bases, those on the hindwings are triangular.
Length: ♂ 42 mm, ♀ 40 mm.

Range: Transition. Widespread in the southern half of B.C.

Field notes: Found around boggy forest ponds and swampy lake bays; favours slightly acidic waters, where it can be surprisingly abundant. Unlike some of our other king skimmers, it often rests on rocks, logs, floating waterlily leaves or the bare ground and can be tame. Many experts place this species in the genus *Ladona*. Flight period: B.C., late May to early September (generally most common early in the summer).

Common Whitetail   *Libellula lydia*

Shows striking differences between the sexes. Male's wings have a broad dark-brown band, and the dark stripe at the wing base is bordered with white on the hindwing; when mature, the male's broad abdomen is covered with bright white pruinescence. Female has brown wingtips, and patches at the base and in the middle of each wing. The sides of abdominal segments on females and immature males have diagonal yellow-white spots. Length: ♂ 45mm, ♀ 41mm.

Range: Southern. Valleys of southern B.C.

Field notes: Inhabits ponds, pools in streams, puddles and quiet corners of lakes. Prefers muddy conditions and tolerates pools trampled by livestock. Typically perches on the ground or low twigs. Males raise their bright abdomens to threaten other males. Many experts place this species in the genus *Plathemis*. Flight period: B.C., early May to late September.

Four-spotted Skimmer *Libellula quadrimaculata*

Grey-brown to yellow-brown, except for the black end of the abdomen. The thorax has yellow and black marks on the sides; the abdomen has narrow yellow side stripes. Each wing has a small dark spot at the midpoint of the front edge, and the hindwings have a dark triangular patch at the base. A golden stripe on the front edge of each wing can be prominent or vague. Length: ♂ ♀ 43mm.

Range: Widespread; also across Eurasia. Throughout B.C. and the southern half of the Yukon.

Field notes: Lives in marshy-edged waters, from mountain and northern bogs and fens to warm alkaline ponds in southern valleys; most common in acidic waters.
One of the earliest dragonflies in the spring.
Flight period: B.C., late April to early October; Yukon, late May to late July.

Eight-spotted Skimmer *Libellula forensis*

Each of the four wings bears two large dark patches; mature males and some females have white patches between the dark ones and near the wing tips, which are clear. The thorax has two pale stripes on the sides, often broken into spots. The abdominal segments have yellow stripes on the sides. Mature males have a thin coating of blue-grey pruinescence on the abdomen and front of the thorax. Length: ♂ 48 mm, ♀ 46 mm.

Range: Montane. Across B.C. in southern valleys.

Field notes: Especially common on the south coast and in the Thompson-Okanagan region. Conspicuous around marshy lakes and ponds at low to medium elevations. Flight period: B.C., early May to late October.

Skimmers

♂ Eight-spotted Skimmer

♂ Twelve-spotted Skimmer

Twelve-spotted Skimmer *Libellula pulchella*

Dark patches at the tip, middle and base of each wing give this species its English name; mature males have white patches between the dark ones and at the bases of hindwings. The body coloration is similar to the Eight-spotted Skimmer's, but the dark wing tips distinguish the Twelve-spotted. Females can be confused with Common Whitetail females where their ranges overlap, but the Twelve-spotted Skimmer is larger and its yellow abdominal stripes are narrower and more continuous. Length: ♂ 51 mm, ♀ 48 mm.

Range: Southern. Southern valleys east of the Coast Mountains.

Field notes: Lives in exposed, nutrient-rich, marshy lakes and ponds especially on alkaline soils. Adult males are aggressively territorial. Common in the Thompson-Okanagan region, but scarce throughout the rest of its B.C. range. Most of its habitat has been drained and filled in the past century. Flight period: B.C., late May to mid September.

Pondhawks *Erythemis*

Most pondhawks range in southern areas and only two similar species live as far north as Canada, one eastern and one western. The genus name, *Erythemis*, means "the red one", referring to several red species in the American tropics; northern species are green or blue.

Western Pondhawk *Erythemis collocata*

Medium-sized and grass-green with clear wings. The face and eyes are also green, though the eyes turn blue with age. Females and young males have a dark stripe on top of the abdomen. Males turn blue with pruinescence as they age, but females remain green. Length: ♂ 42 mm, ♀ 41 mm.

Range: Western. Lowlands of B.C.'s south coast and at the north end of Osoyoos Lake in the southern interior.

Field notes: Lives around ponds and marshy lakes, especially where floating plants occur. It usually perches flat on the ground. Flight period: B.C., mid May to early October.

Blue Dasher *Pachydiplax*

A North American genus that contains only one species.

Blue Dasher *Pachydiplax longipennis*

The thorax has yellow and brown stripes; the base of the hindwing has an orange patch with two dark-brown streaks. Females and young males have brown eyes and a dark brown abdomen with two interrupted yellow stripes on the top. Mature males have a white face and green eyes; the abdomen is thickly coated with pale blue pruinescence, but the thorax is usually just thinly pruinose. Length: ♂ 38 mm, ♀ 36 mm.

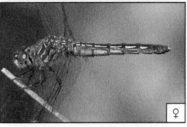

Range: Southern. In B.C., only in the lowlands of the south coast and at the north end of Osoyoos Lake in the southern interior.

Field notes: Abundant across most of the southern half of North America, but restricted to a few areas of southern B.C. Most common on southern Vancouver Island and in the Gulf Islands. Often common at ponds and lakes with abundant vegetation in the water and along the shore. Males defend territories aggressively; both sexes defend feeding perches. They perch, with wings often cocked downward, on stems and twigs from near the ground to high in trees. Flight period: B.C., early June to mid September.

Meadowhawks *Sympetrum*

Small to medium-sized dragonflies that are mostly yellow when young and mostly red when mature; one common species is black. Females are usually yellow or tan, but can be red like males. You can watch most species easily at close range, because the adults are not powerful flyers and perch often. They are frequently abundant around ponds and lakes and adjacent meadows, especially in the late summer and fall. Many species will perch on the ground; *Sympetrum* means "with (or on) the rocks". Species can be difficult to distinguish. Look for the colour of the face, legs and wing veins; the patterns on the sides of the thorax and abdomen; and the details of the genitalia (the male's hamules and the female's vulvar lamina).

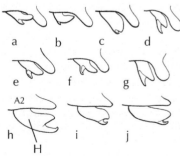

Figure 17. Meadowhawks, male hamules, side view: a, Cardinal; b, Variegated; c, Red-veined; d, Yellow-legged; e, Saffron-winged; f, Western; g, Black; h, Cherry-faced; i, White-faced; j, Striped. A2 = abdominal segment 2; H = hamule.

Variegated Meadowhawk   *Sympetrum corruptum*
The face is tan, but the male's turns orange-red with age. The thorax has two pale stripes on each side, which on mature males shrink to yellow spots at the lower ends. The abdomen is grey-brown and orange or red with white spots on the sides bordered above by a black line. The wings are clear with pink to orange veins. The legs are black, often with yellow sides. Hamule, fig. 17b; vulvar lamina short with shallow lobes. Length: ♂ 40 mm, ♀ 41 mm.

Range: Widespread; also in far eastern Russia. Widespread in southern B.C. to about 52°N.

Field notes: Uncommon in B.C.; mostly in the southern lowlands. Breeds in a variety of waters, often temporary ones, from rich marshes to alkaline ponds and sand-bottomed beach lagoons; usually absent from peatlands. Emergence is long and irregular, and perhaps represents two generations – one in spring and another in late summer. This species wanders widely and some fly southward in the early fall. Flight period: B.C., early May to early October.

Cardinal Meadowhawk   *Sympetrum illotum*
The face is red and the thorax brown-red with two white spots low on each side (the remnants of pale stripes in immatures). The wing bases are orange streaked with dark brown. The legs are red-brown. Male's broad abdomen is brilliant red; female's is duller. Hamule, fig. 17a; vulvar lamina spout-like. Length: ♂ 38 mm, ♀ 37 mm.

Skimmers

(See also the larger photograph on page 4.)

Range: Montane, as far south as Argentina and Chile. Lowlands of B.C.'s south coast to about 50°N.

Field notes: A common and striking resident of lowland coastal ponds and lakes. Males are easy to approach; they return again and again to a favourite twig over the water, perching with wings cocked downward, scarlet abdomen glowing in the sun. Female lays her eggs in tandem with male. Flight period: B.C., mid May to late August (mostly in June and July, which is early for a meadowhawk).

Red-veined Meadowhawk *Sympetrum madidum*

Similar to the Cardinal Meadowhawk, but the male's abdomen is not as broad, the orange wing bases do not have dark streaks and the legs are black. Immatures are grey-brown; males and some females become red and, with age, darken to wine-red. The thorax sides have a pair of white stripes that reduce to spots on mature males. The wing veins are yellow, turning red with age; old ones have brown-tinted wing membranes. The sides of the abdomen bear a white stripe bordered above by black, but these marks fade with age. Hamule, fig. 17c; vulvar lamina a third the length of segment 9, its lobes short and triangular. Length: ♂ ♀ 39 mm.

Range: Western. Widespread in southern B.C. to about 53°N. Probably occurs in northern B.C., because there is one record in the southern Yukon.

Field notes: Uncommon. Develops in a variety of still-water habitats, including marshes, sedge fens and saline grassland ponds, which often dry up in summer. Females lay eggs in water or on the beds of dry pools. Flight period: B.C., late May to late September; Yukon, late June to mid August.

(This photograph is enlarged on the front cover of this book.)

Yellow-legged Meadowhawk *Sympetrum vicinum*
Distinctive for its lack of body markings. Immatures are yellow-brown with unmarked yellow legs; mature males and some females become mostly red with red-brown legs. Hamule, fig. 17d; vulvar lamina strongly extended into a spout (fig. 18a). Length: ♂ 33 mm, ♀ 32 mm.

Range: Southern. Lowlands north to about 50°N.

Field notes: Rare in the southern interior – recorded only from the southern Okanagan Valley and the Kootenay River marshes at Creston. More common on the south coast. Lives in ponds, slow streams and lakes with dense emergent vegetation. While in tandem the female deposits her eggs along the banks in moss or vegetation very close to, or in, the water. The eggs will not hatch until submerged in water. Flight period: B.C., early July to mid November (later than any other species in the province).

Saffron-winged Meadowhawks

Saffron-winged Meadowhawk *Sympetrum costiferum*
The wings have yellow veins and a yellow stripe along their front edges. The stripe is especially obvious on immatures, which are all yellow, except for thin dark lines on the sides of the thorax, black marks on the legs (sometimes the legs are all dark) and a narrow black line on each side of the abdomen. As they age, males and some females turn dark red, the lines on the sides of the thorax disappear and the yellow wing stripe usually fades; other females turn brown. Hamule, fig. 17e; vulvar lamina a short, unlobed trough. Length: ♂ ♀ 36 mm.

Range: Transition. Widespread in B.C.

Field notes: Common, at least in southern lowlands. Inhabits ponds and lakes, especially in the open, including alkaline ponds in grasslands. Flight period: B.C., early June to early November.

Western Meadowhawk   *Sympetrum occidentale*

Distinguishable from others by the wide band of yellow or brown on the basal half of the wings. The yellowish face darkens to brown with age; the yellow-to-brown thorax has dark lines, thickest around the leg bases; the legs are black. The abdomen has black side stripes against brown on immatures, red on mature males and either colour on mature females. Hamule, fig. 17f; vulvar lamina a short, shallow trough.
Length: ♂ 36 mm, ♀ 34 mm.

Range: Western. Widespread in B.C. south of 52°N.

Field notes: Not as common as some of its relatives, but lives in a variety of ponds, marshes and lakes, especially shallow, grassy or reedy places. Foragers and mating pairs often wander well away from water in open country. Some

experts consider it a subspecies of the eastern Band-winged Meadowhawk (*S. semicinctum*). Flight period: B.C., mid June to mid October.

Black Meadowhawk   *Sympetrum danae*

The only meadowhawk in our region with no red on it: mature males are almost all black; females also turn mostly black, but usually not as much as males. Immatures have yellow lines and spots on the sides of the thorax; the abdomen is black with pairs of yellow spots on the top. Hamule, fig. 17g; vulvar lamina spout-like. Length: ♂ 32 mm, ♀ 31 mm.

Range: Northern; also across northern Eurasia. Widespread in B.C. and in the forested parts of the Yukon as far north as the Porcupine River basin.

Field notes: Lives in a wide range of habitats from mountain and northern peatland pools to warm lowland marshes and ponds that dry up in summer. Especially likes peatlands. Flight period: B.C., mid June to late October; Yukon, mid July to mid September.

Cherry-faced Meadowhawk *Sympetrum internum*

Similar to the White-faced Meadowhawk, except for the red face and wing veins. Immatures have a yellowish face and yellow-brown body, both becoming red with maturity. The thorax sides are unmarked and the legs are black. The abdomen has black saw-toothed stripes on the sides. Hamule, fig. 17h; vulvar lamina, fig. 18d. Length: ♂ ♀ 34 mm.

Range: Transition. Widespread throughout B.C. (but not recorded on Vancouver Island) and the southern valleys of the Yukon.

Field notes: Common in slow streams, grassland ponds, cattail marshes and peatland pools. Tolerates many conditions, including cattle-trodden pools and acid and alkaline waters. Sometimes abundant, especially around grassland ponds. At the northern limit of its range in the dry southern valleys of the Yukon, it lives mainly in sedge marshes. Pairs lay eggs onto moist ground while in tandem, often gathering in large congregations. Flight period: B.C., mid June to mid October; Yukon, late June to early September.

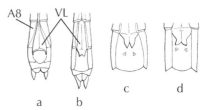

Figure 18. Meadowhawks, female vulvar laminae: a, Yellow-legged; b, Striped; c, White-faced; d, Cherry-faced. A8 = abdominal segment 8; VL = vulvar lamina.

White-faced Meadowhawk — *Sympetrum obtrusum*

Similar to the Cherry-faced Meadowhawk, except for the white face and dark wing veins. Immatures are yellow to yellow-brown that becomes red, especially on the abdomen, as males and some females mature. The thorax is unmarked, but the abdomen has black saw-toothed side stripes. The legs are black. Hamule, fig. 17i; vulvar lamina, fig. 18c. Length: ♂ 33 mm, ♀ 31 mm.

Range: Transition. Widespread in B.C. as far north as Fort Nelson in the east and the Skeena River in the west.

Field notes: Probably the most common meadowhawk in most of southern B.C.; rare in the far northeast. Inhabits a variety of ponds, marshes and peatlands in valleys and mountains. More than most meadowhawks, it can be common in peatlands, and is often more associated with forested areas than the Cherry-faced, which reaches its greatest abundance in grasslands and open terrain. Flight period: B.C., mid June to mid October.

Striped Meadowhawk — *Sympetrum pallipes*

Similar to the White-faced Meadowhawk, but has a yellow face and a pair of yellow-white stripes on the sides of the thorax and usually a smaller pair on top of the thorax. Immatures are yellow to yellow-brown, becoming red, especially on the abdomen, as males and some females mature. The legs are brown, often with pale markings, usually becoming all black, and the wings are laced with brown veins. The sides of the abdomen have saw-toothed black stripes. Hamule, fig. 17j; vulvar lamina, fig. 18b. Length: ♂ ♀ 35 mm.

Range: Western. In B.C., widespread south of 54°N.

Field notes: Common, especially in southern valleys and on the south coast. Found at marshy lakes and around many kinds of ponds, from acidic peatland waters to temporary saline pools. Flight period: B.C., early June to early November.

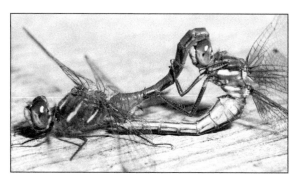

Striped Meadowhawks

Whitefaces *Leucorrhinia*

Small black dragonflies with white faces. The thorax and abdomen are usually marked, males with red and females with yellow (or sometimes red); some pruinescence develops with age. The hindwings have a distinctive small, triangular dark patch at the base and the legs are black. Five of the six species in our region are northern in distribution and most prevalent in the mountains or in the north around the marshy shores of lakes in the late spring or early summer. The Dot-tailed Whiteface is different, preferring cattail marshes and ponds in warm valley bottoms. Whitefaces perch on the ground, logs, lily pads or low vegetation. Males usually hover nearby while females lay eggs. Species can be tricky to separate; look for size, the colour pattern on the abdomen, and the details of the female's vulvar lamina and the male's hamules.

Boreal Whiteface
Leucorrhinia borealis

Similar to the more common Hudsonian Whiteface, but larger and with bigger abdominal spots, including one on segment 8 in males. The spot on segment 7 reaches the end of the segment. The yellow markings of immatures turn red with age. Hamule, fig. 20a; vulvar lamina, fig. 19a.
Length: ♂ 39 mm, ♀ 37 mm.

Range: Northern, but only west of Hudson Bay. Widespread in B.C. east of the Coast Mountains; north in the Yukon to the Porcupine River basin.

Field notes: Our largest whiteface, uncommon in the southern part of its range, but more abundant in northern B.C. (especially east of the

Rockies) and the southern Yukon. Prefers deep sedge marshes, but lives in a variety of fens and ponds. Primarily a species of marshes at the northern reaches of the Great Plains, where it often swarms in large numbers. The flight season is early and short. Flight period: B.C., mid May to early August; Yukon, late May to late July.

Hudsonian Whiteface   *Leucorrhinia hudsonica*
Similar to the Boreal Whiteface, but smaller and with less red or yellow on its abdomen. Male has strong red spots on top of segments 1 to 7, the one on 7 longer than its width; female's spots can be yellow or red. Hamule, fig. 20b; vulvar lamina, fig. 19b. Length: ♂ 29 mm, ♀ 28 mm.

Range: Northern. Widespread in B.C. and the Yukon.

Field notes: The most widespread whiteface in our region and the most common skimmer in the Yukon. Most abundant at boggy lakeshores and peatlands in the mountains and the north. Flight period: B.C., late April to early September (rare after mid August); Yukon, late May to late August (rare after mid July).

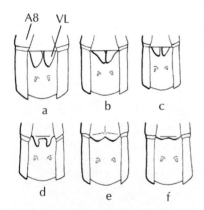

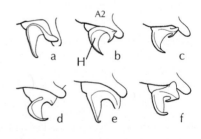

Figure 19. Whitefaces, female vulvar laminae, view from below:
a, Boreal; b, Hudsonian; c, Canada;
d, Dot-tailed; e, Red-waisted;
f, Crimson-ringed.
A8 = abdominal segment 8;
VL = vulvar lamina.

Figure 20. Whitefaces, male hamules, side view: a, Boreal;
b, Hudsonian; c, Canada;
d, Red-waisted; e, Crimson-ringed;
f, Dot-tailed.
A2 = abdominal segment 2;
H = hamule.

Whitefaces

Hudsonian Whiteface

Canada Whiteface *Leucorrhinia patricia*

Tiny and slender with a mostly black abdomen. The top of the male's abdomen is usually unmarked behind segment 3; if any marks are present, they are red streaks on segments 1 to 5. Females have larger yellow marks on segments 1 to 6 (and sometimes a streak on segment 7). Hamule, fig. 20c; vulvar lamina, fig. 19c. Length: ♂ 27 mm, ♀ 25 mm.

Range: Northern. Central and northern B.C., but not found west of the Coast Mountains. In the Yukon, as far north as the Porcupine River basin.

Field notes: Flies alongside its close relative, the larger and redder Hudsonian Whiteface, but is much less common. Rare in the southern parts of its range. Restricted to peatland waters with mats of aquatic moss floating on or near the surface. Flight period: B.C., mid June to late August; Yukon, mid June to early August.

Crimson-ringed Whiteface *Leucorrhinia glacialis*

Can be confused with the Red-waisted Whiteface, but its wings have two rows of cells (where shown in fig. 21a). Male's abdomen is mostly black, with red only at the base and without spots on top of the middle segments (though some have thin streaks). Young females have yellow marks on top of segments 1 to 7, sometimes turning red with maturity. Hamule, fig. 20e; vulvar lamina, fig. 19f. Length: ♂ 36 mm, ♀ 35 mm.

Range: Transition. Widespread in southern and central B.C.

Field notes: Lives around marshy lakes and ponds, especially peaty ones, in forests and mountains. Can be abundant in these places. Flight period: B.C., mid May to early September.

Crimson-ringed Whiteface

Red-waisted Whiteface *Leucorrhinia proxima*

Similar to the Crimson-ringed Whiteface, but all or some of the wings have one row of cells (where shown in fig. 21b). Male's abdomen is mostly black, with red only at the base and without spots on top of the middle segments (though some have thin streaks). Young females have yellow marks on top of segments 1 to 7, sometimes turning red with maturity. Hamule, fig. 20d; vulvar lamina, fig. 19e.
Length: ♂ 36 mm, ♀ 35 mm.

Range: Northern. Widespread in B.C and the main river valleys of the Yukon.

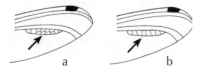

Figure 21. Comparing the ends of the forewings of (a) Crimson-ringed and (b) Red-waisted whitefaces.

Red-waisted Whitefaces

Field notes: Common in B.C., less so in the Yukon; after the Hudsonian, this is the most widely distributed whiteface in our region. Often abundant around mountain and northern lakes, ponds and peatlands. In the Yukon, it prefers sedge marshes and ponds. Flight period: B.C., mid May to early September; Yukon, early June to mid August.

Dot-tailed Whiteface   *Leucorrhinia intacta*
Distinguished by a yellow spot – actually a pair of dots – on abdominal segment 7. On mature males, this spot stands out against the dark abdomen. Immature males, many mature females and very few mature males have yellow on other segments; occasionally, these marks may be red on mature females. The male's lower appendage is forked, the tips widespread. Hamule, fig. 20f; vulvar lamina, fig. 19d.
Length: ♂ ♀ 34mm.

Range: Transition. In B.C., mostly found in southern valleys; east of the Rocky Mountains it occurs as far north as the Fort Nelson area.

Field notes: Most at home in warm, non-acidic lowland waters, unlike the other whitefaces in our region. Common in ponds (it likes organically rich places, such as farmyard ponds) and the marshy corners of lakes in the south. Flight period: B.C., early May to late August.

Dot-tailed Whiteface

Saddlebag Gliders *Tramea*

These medium-sized to large dragonflies come in red, brown or black. Their bodies and wings are similar in shape to those of the rainpool gliders. The English name comes from the dark band or patch at the base of the hindwing that suggests a saddlebag. Egg laying involves a sort of dance – the male repeatedly releases his partner to dip her eggs into the water, then takes her up again. Most species are southern in distribution.

Black Saddlebags *Tramea lacerata*

The body is black with white or yellow spots on top of most abdominal segments; the spots darken with age, but remain on segment 7. The base of the hindwing has a broad black band. Length: ♂ ♀ 50 mm.

Range: Southern; also in Hawaii. A wanderer to southern Vancouver Island, but might also be expected in the Lower Mainland and the warm valleys of the southern interior, which are nearer to the closest breeding areas in central Washington.

Field notes: First appeared in the Victoria area in 1995, then more often in 1996, and a few times since, but no one has observed it breeding there. Common south of B.C. Lives around ponds and lakes, but ranges widely. Usually seen flying and gliding; when perching, it usually hangs vertically or obliquely. Flight period: B.C., mid June to mid September.

Rainpool Gliders *Pantala*

Medium-sized yellow to brown dragonflies with long wings; the hindwing is especially broad at the base. Strong flyers, they can soar and glide, and often migrate long distances. Their wandering is aided by their ability to colonize temporary pools; the larvae can grow rapidly. Only one species has been recorded in B.C., but some are common farther south. The Wandering Glider (*Pantala flavescens*), the only dragonfly that ranges throughout all the world's tropics, is famous for its flights across oceans. It may travel this far north and is worth looking for, though it has not yet been recorded in B.C.

Spot-winged Glider *Pantala hymenaea*

Named for the distinctive round brown spot on the base of the broad hindwing. The body is brown and grey; the face is yellow to orange, turning red with age, at least in males. The closely related Wandering Glider is yellow-orange with clear wings. Length: ♂ 46 mm, ♀ 47 mm.

Range: Southern. A visitor to southwestern B.C.

Field notes: I had a good look at this distinctive species as it hunted over my lawn in Victoria during two days in July 1988. These are the only records in B.C., but this glider probably wanders here more often in July and August. Global warming may result in more sightings here. If it begins to breed in the province, this glider will frequent ponds, including temporary ones. The larvae will probably not overwinter here; adults will have migrated north in the spring and their offspring, emerging later in the summer, will fly south.

Glossary

Abdomen The posterior major division of an insect body (with the head and thorax). The dragonfly abdomen is long and slender and has ten segments.
Anal loop A group of cells in the wing venation of the hindwing; its shape (from semicircular to foot-shaped) is useful in the identification of dragonflies.
Anisoptera The true dragonflies, as distinct from the damselflies; one of the three suborders of the Odonata.
Appendages The projecting structures at the tip of the abdomen. All dragonflies, male and female, have a pair of upper appendages; damselfly males have a pair of lower appendages, but true dragonflies males have only one. The male uses its appendages to clasp the female during mating. The structure of the appendages is important in species identification.
Bog An acidic, low-nutrient peatland dominated by sphagnum mosses; water comes only from rain and snow, not from ground water.
Burrower The larvae of many clubtails and spiketails that dig under sand, mud or rocks and await their prey. The larvae of some petaltails make true burrows in mucky soil.
Clasper The larvae of most damselflies and darners have clasping legs that help them climb about in aquatic vegetation to hunt prey. The term, "climber", has also been used.
Cuticle The outer covering (skin) of an insect, shed during moulting.
Damselfly A dragonfly of the suborder Zygoptera.
Dragonfly A member of the order Odonata. A member of the suborder Anisoptera, to distinguish it from a damselfly; the Anisoptera can be called true dragonflies to allow the use of dragonfly for the Odonata as a whole. See Odonate.
Emergence When a metamorphosed dragonfly, still in larval cuticle, leaves the water and breaks through the cuticle as an adult. It emerges from both the water and the exuvia.
Emergent vegetation Plants rooted underwater and growing up above the water's surface.

Glossary

Exuvia (plural: exuviae)   The cast-off cuticle from any larval moult, including the metamorphosis into the adult stage.
Fen   A peatland affected by flowing groundwater and, thus, richer and less acidic than a bog. A fen is dominated by sedges, grasses and non-sphagnum mosses
Flyer   Dragonflies, such as darners, spiketails, river cruisers and emeralds, that fly most of the time they are active, seldom perching when hunting or looking for mates.
Flight period   The period that adults of a particular species can be found; usually longer than the lifespan of any individual.
Genus (plural: genera)   A group of closely related species.
Gills   Larval structures used in respiration. Damselflies have three leaf-like gills at the end of the abdomen, which they also use for swimming (sometimes called caudal gills or caudal lamellae). The gills of our other dragonflies (Anisoptera) are inside the rectum.
Guarding   The action of a male protecting his mate during egg-laying; the male may continue to grasp the female or he may hover nearby, chasing off other males.
Hamules   Part of a male dragonfly's secondary genitalia, they are paired structures that project from a pocket on the underside of abdominal segments 2 and 3 and hold the female's abdomen in place during mating.
Immature   An adult older than a teneral, but still without adult coloration.
Labium (plural: labia)   The rear segment of the insect mouthparts, usually acting like a lower lip; in dragonfly larvae, it is expanded into a hinged grasping organ tipped with pincers that can be thrust out to capture prey.
Larva (plural: larvae)   The developmental form of an insect between egg and adult.
Mandibles   The chewing mouthparts; jaws.
Mature adult   A fully coloured adult of reproductive age.
Membranule   A small opaque area at the base of the hindwing of most Anisoptera. It may be pale or dark, but is not considered a wing spot.
Metamorphose   Transform from a larva into an adult dragonfly.
Moult   The process of shedding the cuticle. Most dragonfly larvae moult 10 to14 times before becoming an adult.
Odonata   In animal classification, the order of insects to which dragonflies belong. *Odonata* means "toothed jaws", referring to the strong chewing mandibles.
Odonate   A member of the order Odonata. An anglicized version of Odonata often used as an inclusive name for Zygoptera and Anisoptera.
Ovipositor   A structure consisting of blades and sheaths at the end of the female's abdomen in damselflies, darners and petaltails that is

Glossary

used to lay eggs in plants; in spiketails it is simpler and looks like a spike.

Peatland   Aquatic habitat where plant decomposition is so slow that peat accumulates; most common in places where water is generally cold and low in nutrients.

Percher   A dragonfly that, when active, usually makes short flights from a perch. Damselflies and most skimmers are perchers.

Prothorax   The first (front) segment of the thorax bearing the front pair of legs and the head.

Pruinescence   A white, grey or pale-blue powdery bloom on the surface of the cuticle that develops on some dragonflies as they mature.

Pruinose   Covered with pruinescence.

Pterostigma   Thick, coloured cell on the front edge of the wing near the tip.

Secondary genitalia   The structures on the underside of a male's abdominal segments 2 and 3, for transferring sperm to females.

Sedge   A grasslike plant of the genus *Carex* with triangular, solid stems; sedges usually grow in wet areas.

Sprawler   A type of dragonfly larva that hunts by waiting for prey on the bottom sediments or debris (often camouflaged with algae and silt) or on submerged plants. It is called a sprawler because it waits with legs extended. Most emeralds and river cruisers and many skimmers are sprawlers.

Tandem   A position before or after mating, when a male grasps a female with his abdominal appendages. A male damselfly holds the female's prothorax; the male of other dragonflies grasps the female's head. Some pairs remain in tandem after mating, while the female is laying eggs.

Teneral   A newly emerged adult, soft, weak and pale.

Thorax   The middle major division of an insect body (between the head and the abdomen). The thorax has three segments and bears the wings and legs.

Vulvar lamina   The flap-like or spout-like structure (evolved from an ovipositor) projecting from the underside of segment 8 in female dragonflies that do not lay eggs with an ovipositor.

Vulvar spine   A sharp projection on the underside of abdominal segment 8 in some female pond damsels, notably American Bluets.

Wheel position   The mating position for dragonflies: once a pair is in tandem, the female, if ready to mate, will lock the tip of her abdomen in the male's secondary genitalia at the base of his abdomen so that he can transfer sperm. The mating pair's bodies form a circle or wheel shape.

Zygoptera   The suborder of Odonata containing the damselflies.

Recommended Reading

The Dragonflies of British Columbia by R.A. Cannings and K.M. Stuart, 1977 (British Columbia Provincial Museum [Royal B.C. Museum]). This handbook is long out of print and out of date, but if you can find a copy, it is still a valuable reference for dragonflies in B.C.

Dragonflies of Washington by Dennis Paulson, 1999 (Seattle Audubon Society). A fine little booklet (only 32 pages) that describes, with pictures, many of the species that occur in our region.

Dragonflies Through Binoculars by S.W. Dunkle, 2000 (Oxford University Press). This excellent field guide covers all the true dragonflies (Anisoptera) in North America.

Dragonflies of the World by J. Silsby, 2001 (Smithsonian Institution Press). A large colour book containing all sorts of information on the Odonata around the globe.

More technical (and expensive) but important books for the serious student:

The Odonata of Canada and Alaska. Volume 1 (1953) and volume 2 (1958) by E.M. Walker; volume 3 (1975) by E.M. Walker and P.S. Corbet (University of Toronto Press).

Damselflies of North America by M.J. Westfall Jr and M.L. May, 1996 (Scientific Publishers, Gainesville, Florida).

Dragonflies of North America by J.G. Needham, M.J. Westfall Jr. and M.L. May, 1999 (Scientific Publishers, Gainesville, Florida).

Dragonflies: Behavior and Ecology of Odonata by P.S. Corbet, 1999 (Cornell University Press). A wonderfully detailed book.

The internet is alive with information:

http://www.ups.edu/biology/museum/UPSdragonflies.html
   The University of Puget Sound's site deals with Washington Odonata. It is relevant to B.C. dragonflies and has much to offer, especially high-resolution scans of live dragonfly specimens. A related site on collecting dragonflies is:
   http://www.ups.edu/biology/museum/Odcollecting.html
http://www.afn.org/~iori/
   The site of the International Odonata Research Institute offers

information on collecting and research (including collectors' guidelines), books for sale and links to other dragonfly sites.
http://www.ent.orst.edu/ore_dfly/links.html
The Oregon dragonfly and damselfly survey has links to many dragonfly sites around the world.
http://www.royalbcmuseum.bc.ca/research-collectionsdept/nat-hist/section/entomology.html
The Royal B.C. Museum Entomology site has local information, photos and dragonfly reports. It also links to:
http://livinglandscapes.bc.ca/www_dragon/toc.html
Living Landscapes project reports on dragonflies.

Acknowledgements

Thanks to all my companions in dragonfly study, especially Syd and Dick Cannings, Dennis Paulson, Leah Ramsay, Andrew Harcombe, Gord Hutchings, Rex Kenner and John Acorn; they also helped in the preparation of this book. Over the years, many others provided the information that is distilled here.

This book owes much to Dennis Paulson's *Dragonflies of Washington* and his University of Puget Sound web site. I thank Dennis for permission to use some of his material, especially the specimen scans and the identification key to the dragonfly families, which I modified slightly.

The photographs are a major component of the book, and I thank the people listed below who provided images. In particular, I wish to remember my friend, George Doerksen, who died while doing what he loved best, taking pictures of dragonflies; many of the finest photographs here are his.

Interior photographs by George Doerksen (© RBCM), except for those on pages:
10 (all), 11A, 13BR, 14-15 (all), 16C-R, 17L, 31B, 37B, 45A, 47-48 (all), 49A, 50, 54B, 58, 61A, 63A-B, 65A-B, 67L, 69A-B, 75R, 76A, 78B,
    85A-B by Robert A. Cannings (© Robert A. Cannings or RBCM)
43, 51R, 55B, 72A, 75L, 77B, 79A © by Ian Lane
22 (fig. 1), 28B, 55A, 56B, 88, 89 © by Dennis Paulson
37M, 39A, 64, 76B © by Blair Nikula
7L, 9L, 9R by Robert A. Cannings and Brent Cooke (© RBCM)
36 © by John Acorn
16L © by Richard Cannings
17R © by Sydney Cannings
68 © by Sidney Dunkle
28A © by Netta Smith
(A = above; M = middle; B = below; L = left; R = right; C = centre.)

Index to Species

(Page numbers in **bold** locate descriptions and those in *italics* locate additional photographs.)

Aeshna californica 51
   canadensis 46
   constricta 50
   eremita 45
   interrupta 45
   juncea 47
   multicolor 52
   palmata 49
   septentrionalis 48
   sitchensis 48
   subarctica 48
   tuberculifera 46
   umbrosa 50
Amphiagrion abbreviatum 33
Anax junius 52
Argia emma 33
   vivida 32
Baskettail, Beaverpond 14, **60**
   Spiny 14, **59**
Blue Dasher 15, **75-76**
Bluet, Alkali 16, **37**
   Boreal 15, 16, **38**
   Familiar **37**
   Hagen's 15, **39**
   Marsh 14, 15, **39**
   Northern 15, **38**
   Prairie 34, **36**
   Subarctic 16, 34, **35**
   Taiga 15, 16, 34, **35**
   Tule 14, 15, **37**
Calopteryx aequabilis 28
Clubtail, Olive 17, **56-57**
   Pronghorn 14, **56**
Coenagrion angulatum 36
   interrogatum 35
   resolutum 35
Cordulegaster dorsalis 57
Cordulia aenea 70
   shurtleffii 70
Corporal, Chalk-fronted 14, **71-72**
Dancer, Emma's *12*, 14, 17, 32, **33**
   Vivid 17, **32**
Darner, Azure 16, 48, **48-49**
   Black-tipped 14, 16, **46-47**
   Blue-eyed 14, 15, **52**
   California 15, **51**
   Canada 6, *11*, 14, 15, 45, **46**
   Common Green 13, 15, **52-53**
   Lake **45**
   Lance-tipped 15, *22*, **50-51**
   Paddle-tailed *8*, *13*, 14, 15, **49**, 50
   Sedge 15, **47**, 48
   Shadow 14, 17, **50**
   Subarctic 16, **48**
   Variable *10*, *12*, 15, **45-46**
   Zigzag 16, **48**, 63
Emerald, American 14, 24, **70**
   Brush-tipped 17, **61**
   Delicate 16, 60, **63**
   Downy 70
   Forcipate 17, **64**
   Hudsonian 15, 60, 66, 67, **68**, 69
   Kennedy's **63**
   Lake 14, 24, 60, 67, **69**
   Mountain 15, 16, **64**, 65
   Muskeg 16, 60, 64, 65, **66**
   Ocellated 17, **61**
   Quebec 16, 66, **68**
   Ringed 14, 60, 66, **67**, 68, 69
   Treeline 16, 26, 66, **67**, 68
   Whitehouse's 16, 60, **64-65**, 66, 68
Enallagma boreale 38
   carunculatum 37
   civile 37
   clausum 37
   cyathigerum 38
   ebrium 39
   hageni 39
Epitheca canis 60
   spinigera 59
Erythemis collocata 75
Forktail, Pacific *9*, 15, 39, 40, **41**, 42
   Plains 17, **41-42**
   Swift **42**
   Western 15, 39, **40**, 42
Glider, Spot-winged **89**
   Wandering 89
Gomphus graslinellus 56

Index

Grappletail **55**
Ischnura cervula 41
   damula 41
   erratica 42
   perparva 40
Jewelwing, River 17, **28**
Lestes congener 29
   disjunctus 31
   dryas 30
   forcipatus 31
   unguiculatus 30
Leucorrhinia borealis 83
   glacialis 86
   hudsonica 84
   intacta 87
   patricia 85
   proxima 86
Libellula forensis 73
   julia 71
   lydia 72
   pulchella 74
   quadrimaculata 73
Macromia magnifica 58
   rickeri 58
Meadowhawk, Band-winged 80
   Black 15, 16, **80-81**
   Cardinal 4, 5, *13*, 15, **77-78**, 78
   Cherry-faced 15, **81**, 82
   Red-veined *11*, 15, **78**
   Saffron-winged 15, 16, **79-80**
   Striped 15, **82**, *83*
   Variegated 15, 16, **77**
   Western 15, **80**
   White-faced 15, 81, **82**
   Yellow-legged 14, 24, **79**
Nehalennia irene 34
Octogomphus specularis 55
Ophiogomphus colubrinus 54
   occidentis 55
   severus 54
Pachydiplax longipennis 75
Pantala flavescens 89
   hymenaea 89
Petaltail, Black 17, **43**
Pondhawk, Western 15, **75**
Red Damsel, Western 7, 17, **33**
River Cruiser, Western 14, 17, **58**
Saddlebags, Black **88**

Skimmer, Eight-spotted 15, **73**, 74
   Four-spotted 7, 9, 14, 15, **73**
   Twelve-spotted 15, **74**
Snaketail, Boreal 17, **54**, 55
   Pale 14, 17, 24, **54**
   Sinuous 17, **55**
Somatochlora albicincta 67
   brevicincta 68
   cingulata 69
   forcipata 64
   franklini 63
   hudsonica 68
   kennedyi 63
   minor 61
   sahlbergi 66
   semicircularis 64
   septentrionalis 66
   walshii 61
   whitehousei 64
Spiketail, Pacific 17, **57**
Spreadwing, Common *13*, 15, 16, 30, **31**
   Emerald 15, **30**, 31
   Lyre-tipped 15, 16, **30**, 31
   Spotted 15, 16, **29**, 31
   Sweetflag 15, 16, **31**
Sprite, Sedge 15, 16, **34**
Stylurus olivaceus 56
Sympetrum corruptum 77
   costiferum 79
   danae 80
   illotum 77
   internum 81
   madidum 78
   obtrusum 82
   occidentale 80
   pallipes 82
   semicinctum 80
   vicinum 79
Tanypteryx hageni 43
Tramea lacerata 88
Whiteface, Boreal 15, **83-84**, 84
   Canada 16, **85**
   Crimson-ringed 14, **86**
   Dot-tailed 15, 83, **87**
   Hudsonian 14, 15, 16, 83, **84**, 85
   Red-waisted 14, 86, **86-87**
Whitetail, Common 15, **72**, 74